The Absolute Pseudo Truths To The Origins Of Life

The Absolute Pseudo Truths To The Origins Of Life

Dr. Hagen Daahs &
Dr. Ima Fükuppe

Library of Congress Control Number:		2011911758
ISBN:	Hardcover	978-1-4653-3471-8
	Softcover	978-1-4653-3470-1
	Ebook	978-1-4653-3472-5

This book was printed in the United States of America.

To order additional copies of this book, contact:
Xlibris Corporation
1-888-795-4274
www.Xlibris.com
Orders@Xlibris.com
82908

Contents

FORWARD

· · · · · · · · · · ·

INCREASINGLY WE FIND OURSELVES IN a world of religious bias and cultural wars being waged in our political and social structures. The ever shrinking world we share is undergoing a major metamorphosis not seen in many decades with financial uncertainties, climate issues, philosophical challenges, new leadership, political and religious turmoil, etc. As within all ages of uncertainties, we as a people tend to undertake a renewed search for the true meaning within each of our lives. It is the uncertainties of life during this age of dynamic change that has caused so many of us to search for the deeper, absolute sense of truth. We all spend our lifetime in search for the true meaning of life. It is why we, the authors of this book, felt compelled to conduct so much personal research and to document our findings in this pseudoscientific book on such a controversial subject as the intelligence of our creator. We wanted to contribute our knowledge and findings to help in finding the true path forward from the morass of conflicting ideas and the controversy of creationism versus evolution and to elevate this discussion to a new plateau of absolute truth. We believe that once we sat upon the plateau of absolute truth we could gaze down upon creation and all agree on the absolute truth to the meaning of life.

When we first heard of the philosophy (or is it a theory?) called "Intelligent Design" we felt compelled to launch a scientific investigation of this philosophy and discovered that it had a huge number of believers. We could not help but wonder how so many people could believe in a mere postulation that did not appear to have any basis either in plain facts or in science. These persons stated that they truly believe that since evolution was a "theory" it has no validity as if "scientific theories" have no basis in actual data or fact and are, therefore, invalid. We questioned these persons to see if they believed in the "Theory of Gravity". The results were as follows:

- 26% said no because it is "only a theory" and was not scientifically proven.
- 34% said no because it does not apply in their lives
- 16% said it was a liberal, fascist plot for the government to take over their lives.
- 10% had no opinion.
- 44% said they used it every day and were quite pleased with the "Theory."

Those persons believing in Intelligent Design totally ignore the fact that evolution, even under their rationale would have as much validity as any other "belief." Still more unsettling was their stated belief that religious faith in creation had an equal role to play in the realization of the human species and that science was not a valid basis for postulating the origins of life. Many share the belief that an all knowing, all powerful God created all of existence and did so based on clear intent. However, the more unemotional thought we gave to the idea of Intelligent Design, the more we began to realize that, either, there had to be some substance to that lay hidden beneath a layer of obvious "faith" or that there were a great many not very intelligent people in this world who are, in effect, incapable of independent critical thought. Why else would rational human beings posit such a theory? Armed with this new fervor we decided to explore the idea of Intelligent Design and to ascertain the rationality upon which we are sure it must be based. We intended to approach this quest with all the scientific fervor and discipline that it deserved. We came to realize that the entire idea of Intelligent Design had meaning to Christians throughout the entire world, including those living in America, and that we owed this community the unbiased results of our pseudoscientific analysis. This tome is our journey through the morass of thought known as "intelligent design." The absolute meaning of life is a quest which we have, in researching this book, also discovered the answer to. However, we will not reveal it in this book as we believe we can make more money by revealing it in our second pseudoscientific scientific tome, "The Absolute Truth to The Meaning of Life". Hence, please wait for the sequel and all the answers will be provided, a case for "intelligence by design." We have, in researching the hypothesis of Intelligent Design, analyzed the archives of the Vatican library as well as interviewing many religious scholars from around the world and other places to ensure we had no biases in our book. Of the 30,000 manuscripts contained in the Vatican library none were found to reference any "intelligence" at all. Several were found that clearly indicated the universe rotated around the Earth once every 1750 years and the year 2012 was so unique that it would result in a universal alignment of the constellation peniserectus with a very large black hole. Many believe that this alignment of peniserectus with a black hole could have devasting consequences at midnight on December 21, 2012. Dr. Daahs is continuing to research the plethora of factoids which exist in this collection of short stories.

Dr. Fükuppe and Dr. Daahs strongly believe that most people tend to associate their religious beliefs with religion during acts of sex and this is one of the key reasons we have linked the two in this book. Dr. Fükuppe has studied the phenomena wherein persons of either sex will often have a religious moment relating to their god during climax. Her studies were conducted during her videotaping of various sex acts and revealed that 58% of men cried out "oh my god" at the moment of ejaculation and 23% were very specific citing "Jesus, oh Jesus". This study also revealed that 84% of women cried out "for god's sake, don't stop" during climax. 13% of men cried out "oh yeah bitch" which clearly identified them as atheists. 6% of women cried out "oh, oh, oh" identifying themselves as Mormons. Over 34% of women cried "fuck me harder" during climax indicating the richness of the religious experience they were undergoing. 56% of persons using marijuana cried "oh . . . my . . . goooooood." It appeared that their experiences were of longer duration than the others. Scientists have shown that time is not fixed but is a variable and these results from marijuana users support such a hypothesis.

This book is not always continuous in its presentation of thoughts to consider. That is because we very poor writers and because there is so much to ponder. Never the less, this book is an authentic novel based on real pseudo findings and experiments. Bear with us, dear reader.

The authors of this work are world renowned scholars:

1. Dr. Hagen Daahs is 55 years old and has not graduated from Yale or Harvard. He has a doctorate degree in Metaphysical Philosophy from the University of Salem. He is an unassuming person of 5'4" height, which he constantly compensates for by standing on his toes rather awkwardly, and he sports a full

facial beard which is predominantly gray. He has lost most of his hair except for the pronounced chest and pubic hair which is extremely curly, but alas also gray. He most frequently wears plaid sport coats and has food spills on his corduroy trousers and hairs growing from his nose. He wears horn rimmed glasses with tri-focal lenses. He weighs a svelte 145 pounds and wears sandals almost year round. His research papers have earned him global recognition including the Helsinki Medal for Pseudo History for his research into the Mysteries of Multiple Conflicting Historical Records.

2. Dr. Ima Fükuppe (foo-cup) is a striking Germanic/Swedish blonde with piercing steel blue eyes, 6'2" tall, and 135 pounds. Her measurements are 42D" (106.68 cm for the European readers), 28" (71.1 cm) and 34" (86.4 cm) respectfully. She typically wears knee length black leather boots and a blouse open to the waist. She does not wear any brassieres as she finds these restrictive of her right to choose. She can often be seen on campus with her riding crop at her side. She graduated Summa w Cum Loudly from the University of Helgasmorgod earning a Doctorate degree in Bizarre Religious and Medieval Practices. She has a minor degree in Perverse Sexual Practices. She is internationally recognized for her seminal work "Finding God through Sex."

CHAPTER 1

. . . .

The perfect life or,
what the hell was he thinking?

ONE OF THE MOST PERVASIVE fundamental quests we humans have is our desire to understand the basis for creation, to understand where we came from, to understand what we humans really are and what our true purpose on this planet Earth is meant to be. This question has been asked many times in the variations of the form "What is the true meaning of life?" This question is so compelling to humankind that many books and movies have been written purporting to have grasped this elusive Holy Grail (or glass, jug, secret decoder ring, etc.) of understanding. Such classics as "Moses," the holy bible, and the Gospels of Luke, and Johnny and the "The Ten Commandments" are revered among the religious fundamentalists. Among our favorites, the pseudo factual Monty Python version, "The Quest for The Holy Grail" was also produced in an attempt to seek truth, but alas, came up with incomplete answers to these fundamental questions. Still the movie was very humorous and provided many insights that assisted us in our research.

In researching this book we found that an astounding 93% of all humans believe that human existence is a good thing for living people to experience. 94% of the humans on the planet report being less than ecstatic with their life styles and would change many attributes of their worldly existence.

Most of us, whether religious or not, accept the sense that we are, as humans, are incidental players in the reality in which we live. Almost all humans accept the feeling that the universe in which we live is far too complex for us to accept it's complexities as a mere random occurrence. Are we just space matter coalescing in a random fashion creating the very universe we exist in through our collective minds? Indeed creating the very essence of humanity? Are we humans, in reality, simply the result of the random coalescence of just stardust? Nothing could be more unacceptable to the mass of non critical thinkers, whether they are Christian or Muslim or Jewish or Hindu or Hittites or Shintoists or Druids or Scientologists. Apparently man has always sought an easy answer and has chosen to follow others rather than to think for himself. And he continues along this path to this day. Science has measured the age of the universe in hundreds of millions of year timeframes, yet some among us believe that creation should be measured in terms of, well 6000–10,000 years. The reader should note that this is 4–8000 years BC or 22000 years WMIU ("we made it up"). This is very unsettling to the ancestor's of early man since their familial records show communal life in Egypt, Asia and Europe dating to much before 10,000 BC. To many people, it is very difficult to imagine that the entire whole of the universe, all that we know of and all that we don't know of, is not the work of a single god entity who only worked for 6 days (144 hours) and rested on the seventh day. Was this omnipotent being just too exhausted to continue working? Was he a member of a labor union? Imagine what more he might have accomplished if only he had given humanity 24 more hours! Most likely the United God Workers (UGW) union simply prohibited more than 6 days of uninterrupted work without a break in order to preserve jobs for all of the lesser gods. This battle between labor and management exists in our world today and the conservatives in America are doing their part to abolish labor unions, unfortunately thousands of years too late for God to have demonstrated his true creativity.

To the authors, Dr. Daahs and Dr. Fükuppe however, the brilliance of this god's work appears less than compelling. According to the proponents of Intelligent Design, this complexity of life can only be explained by the belief that a supreme being alone could have created such complexities as human life itself as well as the order within the universe of which we are a part. The enormity of the universe in which we live, the complexity of the world we inhabit, the inevitability of body death all require us to look at these questions with a seriousness that transcends both a simple Intelligent Design theory as well as the scientific basis for evolution.

Wanting to establish the veracity of Intelligent Design in our work, we have explored the very limits of scientific theory as it applies to the human body, the cosmos and even to fecal matter. It should be noted, dear reader, that since intelligent design appears to be based on pseudo science we applied the same pseudo scientific rigor to our findings and research. It is essential to realize that the foundations of science require, without exception, that scientists explain their theories, test these theories, retest them and enable independent researchers to verify them without failure in order that the laws of the universe and life itself are truly explainable in repeatable and perhaps mathematical terms. Hence our scientific findings as reported herein, require such verification and we have encouraged new scholars to replicate our findings. We have pursued this scientific quest with much rigor by conducting the actual tests of our theories by experiencing and living actual human lives and we also employed actual pseudo scientists to review the lives of other real people. This research was accomplished on a global basis. It has taken us a lifetime to gather all of the scientific evidence we report on in this book. On the other hand, the religious folk, Christians among them, do not require such "proof." As a preference they shun science as it does not support their belief theory and must, therefore, be incorrect. Only those religious persons with sufficient "faith" can truly understand Intelligent Design. They shun the scientific method but, in fairness, also practice pseudo religious theory through pseudo spokespersons. We, Drs. Daahs and Fükuppe, cannot accept such a poor basis for our conclusions and thus we have written this book which contains years of research in which we attempted to resolve this controversy well beyond the realm of further dispute.

The very complexity of the universe, life itself and the inability of the scientific community to define the laws by which the universe operates is a basis for the Intelligent Design belief. Even those persons who accept the "theory" of gravity in their everyday life point out the fact that scientists have failed to explain precisely what gravity is, and therefore it must not be a reality. On the other hand this conclusion tends to lead some of us to the belief that the world is too complex and imperfect to have been the work of a supreme being, such as an intelligent designer. Especially, a designer that quit working in the seventh day clearly before he achieved perfection. Wouldn't an intelligent designer approach the design of the universe and humankind in a more rational manner? As a starting point, the Intelligent Design proponents believe, for example that man is a personification of the God form. That Jesus is the personification of God, and that he is the all supreme knowing entity. All of this belief based upon blind faith.

Dr. Daahs was particularly concerned with the time of creation. Science has now documented the age of the universe to be between 13.5 and 14 billion years old. "That is considerably greater than the 6–8000 years creationists believe in," he said. "Even with an error of 10.5 billion years that is a great deal of margin in which the god like entity may have created the universe." "Did he start before the 'Big Bang'?" asked Dr. Daahs. "If he did he would have had to compress all that ever was into a tiny dot of material no bigger than a pinhead. In accomplishing this compression he would have crushed all of the angels that could fit on the head of a pin or the entire god like forms into this tiny singularity of matter. "Where did he even stand to get positioned to compress all of this substance into such a tiny dot"? "Where was all this matter and energy before the Big Bang," asked Dr. Daahs. "Was it scattered in an alternate universe? Did this god travel the universe collecting all of the matter and energy in a special container of infinite strength? Did he invent a new super strong material such as Kryptonite? Where did god live before he created the Big Bang? Perhaps he lived in Iowa as that would explain his reluctance to release his birth certificate. Were the fundamentals of life already present when he created Earth? If so why reinvent them? "Perhaps the early life forms lacked breasts? Beautiful large 36" breasts like Dr. Fūkuppe's probably did not exist just a few years after the Big Bang?

Perhaps he started after the big bang took place. The questions then become, did he start with the Earth as a red hot celestial orb of violent volcanic action without pure water and pure air? Did he calm the storms and cool the planet and create water all in the 6 days of creation (144 hours if he took no coffee breaks)? Perhaps he urinated continuously onto the volcano's to cool them down? After all he was god wasn't he? Did he start immediately after the big bang, i.e. in minutes when the universe was an expanding mass of gases and energy and had not yet formed galaxies no less planets? Why did he pick Earth? Mars had plenty of water and an atmosphere? "With less gravity on Mars, wouldn't women's breasts sag less in middle age?" asked Dr. Fūkuppe, "a good reason for choosing Mars over Earth." Although Dr. Daahs points out that the male penis may benefit from the effects of gravity and wonders if Saturn's gravity would not have produced a magnificent sex organ.

Where was god's father in heaven at the time? Was heaven in the clouds of earth eventually to be disturbed by aircraft flying thru heaven?

Or was he on another galaxy taking his time (like 13.5 billion years) instead of grandstanding to accomplish building heaven in seven days. Was he possibly relaxing in an alternate universe? Perhaps the god like entity had a different master plan. Perhaps he did plan the entire universe at the same time.

The proponents of Intelligent Design must posit that we humans are perfect thus reflecting the brilliance of the designer. If in reality we are perfectly designed, then a question might be, "Isn't the human existence also designed to be perfect for all humans? Is there nothing you would change? Perhaps our global database is incorrect, or do you, the reader, vouch that everyone you know is godlike and is living the perfect life? We researched the endless records of known and recorded human existence and looked for the perfect godlike beings. Unfortunately, most of the historic figures we researched are dead. Apparently, even gods, perfect as they may be, die. Or just disappear forever.

Why wouldn't an Intelligent Designer design humans, at least, to live forever? We asked ourselves why is a finite life span Intelligent? If gods are immortal and live forever in another form or in a parallel universe, why not let humans choose an age and be immortal also? We decided that we would choose to remain at age 20 forever. It was a very good year! If he designed us in that fashion he could have imbued us with the same sense of human pleasure from sex but without the bother of procreation since there would exist only a need for replacing the truly stupid people that killed themselves off somehow. For instance, the males that chose to body surf Niagara Falls or those that choose to wrestle alligators in a swamp or those who enjoy Russian roulette with a real gun. These humans would not likely be missed or even noticed that they had disappeared from existence. Evolution working perfectly and strengthening the case for evolution over creationism! Preferably these persons could have been designed to kill each other through duels to the death with likeminded people, thus eliminating inferior humans two at a time instead of having them live forever. Dr. Fükuppe raises another excellent point by indicating how much wiser it would have been if she could have chosen the age at which she could live on to immortality, that age being 72 hours after her breast implants had healed, for instance.

The entire process of aging appears superfluous if we had been intelligently designed. What is the purpose of aging? If the world were not filled with children everywhere constantly evolving into obnoxious adults, then current adults could just live forever without the need for children at all! Our resources would not be strained and we could all vacation for 50 weeks of each 52! Aging does not seem to serve any useful purpose past the age of sexual indulgence. In the modern world, age seems to enable the pursuit of ever increasing amounts of wealth. Is this the result of creationism or evolution? We suspect it is the result of the human genome being changed due to evolution. In a survey conducted by Dr. Daahs, people were asked to choose the age they would choose to remain until eternity, or until forever, whichever came first. The results were as follows:

- 96% of American men said they would like to remain 24 years old.
- 87% of prisoners sentenced to life in prison indicated that eternity is a "very, very long time".
- 89% of American women reported their age would be 32.
- 92% of European women reported their choice of age would be 40.
- 77% of Asian women chose age 13.
- 45% of Chinese men refused to answer.
- 48% of Muslim men kept repeating the phrase "virgins."
- 14% of Christians said they can't wait for eternity to arrive as heaven was preferable to earth.
- 88% of Irish persons questioned said they didn't care.
- Queen Elizabeth said "I don't discuss my personal preferences with commoners."
- 34% did not know what eternity meant.

Dr. Fükuppe indicated that she thought "this data is meaningless."

The evolutionary theorists explain that aging is a process wherein our body cells grow tired of perfect replication. That fact, coupled to a myriad of environmental effects, makes our body age regardless of how well we treat it. As a factual matter if you eat only organically grown vegetable and fruits, exercise every day and treat everyone you meet with love you will likely die. If, on the other hand, you eat animal fats, drink Tequila every day, spew vile hatred while laying on a sofa it is also likely you will die. Dr. Fükuppe has researched the dietary habits of ancient gods and found that almost all ancient peoples and gods ate only organically grown grains and fruits and also ate only open field grazing animals, yet their life spans reportedly were shorter than those of current humans. Apparently god designed them that way. A good argument for evolutionary theory but an absolute indictment for the proponents of intelligent design!

Dr. Goodman[3] has reported in her book, that the reason people age is to make room on earth for more people who will also age and die to make room for still more people who age and die. In her groundbreaking work, Dr. Goodman's research has determined, without any scientific doubt, that almost all humans die, sooner or later. Yet very few actually chose the moment, method or manner of their demise. Some Gods apparently do not die, that is why we often see their images as older and bearded. Since they are immortal they may walk among us even now. This fact is true of the male gods. Most female gods and goddesses like Aphrodite chose to remain voluptuous looking rather than bearded regardless of their age, an inherent trait of all females. It is because they know they will live forever that they want to pretend that they are vulnerable like they designed us humans to be. Imagine the uproar if humans realized that gods were immortal and they deliberately designed us mere humans to live for very short periods. Dr. Fükuppe has questioned the premise that gods do not die asking the question "why would they have designed humans to need sex for procreation if the gods themselves did not have to reproduce? Could it be that the gods practice sex acts just for pleasure. If gods are immortal and procreate, how many omnipotent gods are there?" she asked. "Must be millions by now prolific as they are!" she quipped.

In our research we investigated the many persons that society has judged as perfect, and therefore presumably god like. Apparently there were far too many to list. The list includes the likes of Alexander the

---

[3]   *Dr. Goodman is a world renowned good woman and scholar. She has a degree on Pathological Death and other Fatal Maladies and has studied ancient humans and gods. The most profound factoid resulting from her studies is that it appears every single human or god has died. Some earlier than others, however. Her book entitled, "Death after Life" is the documentation of persons being born and then dying. She is searching for gods that may be immortal and has thus far completed 168,000 face to face interviews without a single god being found.*

not so Great, Buddha, Marilyn Monroe, Caesar, Zeus, Ghandi, George Clooney, Queen Victoria, and Angelina Jolie and on and on. Perhaps among the many many gods that were worshipped on earth by the current humans are many that you would recognize. In reviewing a partial list, both Drs. Daahs and Fükuppe commented that their mental images are all of magnificent bodies and facial characteristics associated with these gods. Are these images then the perfect human form which we were designed to be? What images do you conjure these immortals to possess and, since they are immortal, who do think they pretend to be in today's society? Consider the partial list and the associated images they conjure:

- Aphrodite, Neptune, Venus, Apollo, Odin, Artemis, Athena, Poseidon, Atlas, Baal, Bacchus, Quetzalcoatl, Brahma, Dionysus, Freyr, Hades, Shiva, Hermes, Seti, Iris, Ishtar, Vesta, Isis, Juno, Sol, Krishna, Thor, Loki, Luna, Mars, Zeus, Mercury, Minerva, Vulcan, Mithras.

Their statues and stories reveal the near perfect human form we sought to find. On the other hand, when we tried to correlate this measurement of perfection in actual living beings the data failed to support us. We discovered many anomalies, for instance, we found famous clothing models, certainly god like females, that were born with spaces in their teeth or had skinny legs or no ass. We found movie stars with sagging breasts at age 30 and others who have had their facial skin stretched so tightly that talking produced a "D" major sound. We found male movie "studs" that portrayed lovers in steamy bed scenes who were actually Gay. We found "holy" priests that had forced sex with boys; we actually discovered people who killed other people! We have found people who killed other people simply because they were not of the same religion. While one might expect a few hundred of such deaths, they actually number in the millions! We found innumerable people born with innumerable imperfections. We found, in summary, imperfection everywhere. God did not do such a great job in design. We asked ourselves, "Could it be that god deliberately screwed up the design of certain people?"

However, as the reader must imagine, it is very difficult to compile sufficient research data on the 8.3 billion of earth's population. Try as we did, we could only interview a few of the worlds 8.3 billion people.

Consequently, Dr. Daahs and Dr. Fükuppe welcome your personal data on human perfection. If you know people who are perfect (created in god's perfect image) in every way, or if you know persons who appear not to be designed in a god's image, send us their story and pictures along with names, addresses and a $20.00 check to cover our costs to Perfectionsearch@Fukuppe.net, major credit cards are accepted. We will add their perfection criteria to the Universal Human Perfection Database (UHPD) which will be accessible to all whom wish to verify our research. However, due to the high cost of maintenance, we will not be able to analyze and preserve this critical data without the $20.00 fee. When submitting your data we do not need actual names and addresses, pseudonyms will do just fine. However, be sure to include all of the data, actual or pseudo that makes you believe these people you are submitting qualify them as gods or definitely not gods. Be sure to include their physical attributes as well as their sexual physical attributes and sexual practices that are god like or not. Detail is helpful. Pictures are desirable. Certain persons may be rewarded with a nude picture of Dr. Fükuppe.

In addition to the obviousness of human imperfection in design, a second issue, that of complexity in general, leads us to question the idea of intelligent design even further. The very complexity of the universe, life itself and the inability of the scientific community to define the laws by which the universe operates is a basis for the arguments leading to the Intelligent Design belief. That is, the intelligent design crowd views these complexities as the reason for a belief in a supreme god of creation! On the other hand this conclusion tends to lead some of us to the belief that the world is too complex and imperfect to have been the work of a human like Supreme Being, a truly intelligent designer. The complexity of life within our universe has far too many flaws to have been designed deliberately but reflects an evolutionary path that accepts its many flaws.

Interestingly, a unified theory of the universe using modified string theory now suggests that we may live in a universe of 11 dimensions with 'membranes' separating us from many alternate universes and possible new life forms. This forces us to question the intelligent design belief anew. How could the intelligent designers have designed us to operate in only one universe with such limited tools when there are adjacent universes in which totally different experiences await? For instance, does the species of life in an adjacent universe live forever?

Is that where the gods exist? Imagine another universe in which color may be the sexual satisfaction mechanism. Imagine, "give me red baby" can produce astounding orgasms that last for years! And the duration under the influence of marijuana is truly mind blowing.

Wouldn't an intelligent designer approach the design of the universe and humans in a more rational manner? Why for example, would humans actually be the random collection of stardust if some entity could have made us unique from stardust? Even more preposterous to us is the fact that the separation of genes of humans from other animals such as monkeys is estimated to be approximately 5%. Amazing! The designer must have been getting tired on day five and cut some corners. If we are only separated from monkeys by 5% of our genes, shouldn't the religious leaders be converting monkeys to their religious beliefs and thereby increase the flock of mindless followers? On the other hand, if we are only stardust and randomness it is amazing that we are only 5% different than monkeys! What does the 5% genome change do? For starters the monkeys only come into heat during the menstrual cycle so sex is for procreation only. No climbing into the back seat for a quickie with these mammals. Except that male monkeys are always horny and even have sex with another male if no female is present. What happened to life style choices? The difference in human and monkey males are obviously not contained within the 5% of different genes; Evolution or Intelligent Design? This argument is a winner for the evolutionary camp.

In America the Intelligent Design proponents are predominately Christians for whom the belief that Jesus Christ is the son of God and from this family tree flow's the entire fractal perfectness of the Earth's population. Indeed in the aforementioned 144 hours, this god's relative created the perfect universe. Jesus must have been perfect also. How else would a brilliant designer shape the son of God? Upsetting this wonderful postulation of Christians is the confusing beliefs wherein Muslims believe that Allah, their God, is the only true God and created them in his grand image. Our extensive research has also shown that Judaism has as its fundamental belief that a single entity, as revealed only to Moses, is, of course, the only true God. Many Indians believe that Hinduism and Brahma the only true God. Still other, apparently ignorant people, believe that perhaps Buddha, Joseph Smith or Ra or Zeus are the "true gods." Dr. Daahs has spent his entire lifetime, so far, conducting the search for the <u>one</u> true god, truth being a condition that is necessary less the term "god" became incorrectly applied according to the dictionary. This apparent discrepancy wherein many ignorant people believe that only their version of a true god is true is very disturbing, because there can only be "one" true god. In Italian Mafiosi terminology, the Capo of Capos of Gods! Note: The term "true god" implies that one and only one can be the ultimate god creature. Is it any wonder that man has fought religious wars for over 9,000 years to determine who is the one and only? Wouldn't you expect that the omnipotent one and only he would have been victorious by now? Long live the only true god!

Dr. Daahs has dedicated his life to ascertaining the truth about this conflict. Fortunately Dr. Daahs has been in contact with scholarly archaeologists who have most recently discovered the "Lost Scrolls of Creation[4]" and he has studied each of them in detail.

These scrolls date back long before any record of Christianity and clearly describe an Earth visited by the gods clearly with an intent to create the human race and to provide the basis for intelligent life. To further his quest for knowledge and to probe still deeper into the "Lost Scrolls of Creation[4]" he has also endured months of deep hypnotic

---

[4] *The "Lost Scrolls of Creation" are a collection of recently discovered manuscripts apparently written on human foreskins (predominately Semitic and Egyptian) stitched together after numerous circumcisions. Editor's note: Most of the peoples of this time period did not practice circumcision as they had very dull knife like instruments. The scrolls appear to describe the creation of the universe and of humankind. They are written in the obscure language of Hottits and are believed to have been buried in 543,789 BC by alien visitors to Earth. It should be noted that the term alien refers to entities that are not "born" on earth and arrived here somehow on something. They can be thought of as very similar to Mexicans that seem to just "show up" in America, no one knows how they got here or when or by which method. Nor is it known when, or how, they will depart. They, like the gods, are aliens. The scrolls tell that the aliens introduced circumcision to mankind in order to eliminate odors that they were not fond of after they created humankind. Note: This fact leads us to postulate that the gods were not perfect either. Why design a body part that creates an odor? As a further alien teaching to "waste not, want not" the foreskins were rendered into near perfect writing paper upon which*

were recorded all of these details. In a recent discovery by Dr. Fükuppe, stroking the precious scrolls with two hands while she cradled them in her crotch actually increased their size so they could be read without the aid of reading glasses. "It would have been better if they grew more than 2x in size," she noted, but I guess the gods were not well endowed either and had no respect for women." The exact location of this manuscript and indeed the teachings of it, are known to only a few Hottits scholars. It is believed that their disclosure to the masses of humankind would result in widespread panic among the millions of Intelligent Design creationist believers. The manuscripts survived several millenniums only because they were carefully encased in a golden metallic wrap of unknown material of extreme lightness and strength and buried on a very dry Chinese plateau near the nation that was Tibet. The scrolls have been studied by only a few scholars as the release of such fundamental information could cause the entire world to go to war to prove the teachings false. Each scholar has to contribute a foreskin for scientific purposes just to gain access to the scrolls. As a consequence of this requirement, women scholars are finding it difficult to get male colleagues to contribute on their behalf. Fortunately Dr. Fükuppe was able to convince several donors and consequently she has been able to open the research door to many female scholarly researchers. So shattering are these findings that they reveal the true creation of the universe. According to the Lost Scroll documents, the universe was not created in only 144 hours as many believe. Rather the gods themselves appear confused about this subject as they believed the universe was "always there but somewhere else" ever changing but ever in existence. Many of the gods did not know how old the universe was but all conceded that it was older than a "few hundred million years, give or take a few million years." "It is difficult to judge the age of the universe since we are immortal and don't know how old we actually are" said one very old god. These scrolls further reveal that the creation of humans was decided upon by these gods as an experiment. The scrolls further indicate that female gods were also present but found the creation process for human males wanting and so they left to window shop in the nearby Andromeda Galaxy and spent little time in the actual creation process until they returned. Once returned the male gods left to seek celestial golf courses and a band of female gods alone completed the creation of humankind. They were appalled at the lack of creativity shown by their male counterparts and immediately set upon critical design changes, most notably increasing the tumescent penis to 4x its flaccid state size.

The Lost Scrolls are of such intent interest that the scientific community is devoting extensive research resources to discover their true meaning. The National Security Agency, for instance, has devoted 106 full time super computers to process each word, symbol and scratch to determine the hidden code that could reveal the darkest of intended secrets. To date, this NSA effort has already revealed that cats are the creation of the anti gods, women have far superior temporal brain lobes, and rocks are of little use, the Irish were created for general amusement and that the world will end in the year 2012 unless the humankind creation finds the one true god. The materials with which the scrolls were wrapped for protection is so strong and light of weight, that NASA has begun to build a new spacecraft that will weigh only 4 pounds and be able to seek out the one true god as it travels to the very ends of the universe. Unfortunately this journey will require 300,000,000 years. Applying this material to produce light weight, and therefore highly efficient automobiles, has been deemed to constitute a compromise of national security interests. The Russians and Chinese, having been left out of this endeavor, are mounting a massive spy program using Internet hacker attacks to solicit the details within these scrolls. The governments will not reveal where the scrolls are being stored in fear that Iranian agents working with the North Koreans will also act to steal these precious documents. Both the Koreans and Iranians have a plethora of foreskins to contribute, a factoid that worries many trying to preserve the scrolls. Fortunately only a few Hottits language scholars exist in the entire world and so even loss of the documents will not have an immediate effect on international security. For more on Hottits scholars, see Appendix 1 which is not yet available.

regression to witness firsthand the alien gods that visited upon the earth. The Lost Scrolls are quite revealing in that they appear to have been written by the gods themselves. The Lost Scrolls of Creation has provided him with a deep insight into the reality of the one true god and from clues provided from this work he has established a link under hypnosis and talked with many of the original true gods. A new book, "Dr. Daahs' Conversations with the True Gods" will be published shortly and is dedicated to the creativity and creationism of these gods.

What Dr. Daahs discovered during these deep hypnotic trances, was that several tier 2 gods did try to act in an intelligent manner. They realized that if each god (tier 2) acted according to his own true beliefs, then man would be designed in each of their likenesses. Some were very short. Others very tall; still others were fat and some might claim to be a bit odoriferous while still others appeared to be gay. To avoid this, they formed a committee under the auspices of the one elected omnipotent god, (it is not known if this omnipotent god was the only "true god") so that all of these god deities were equal participants (committee like) in the intelligent design process.

How absolutely clever of the intelligent designer to blend the attributes of so many subordinate individual intelligent designers as the team management method of arriving at the perfect human form. This is the first known record of "Team Management." In essence these deities formed the Human Being Design Board of Directors (HBDBD, (if the US Government were to classify the organization for tax purposes), and were able to compromise on what a true human god like creature should be. (Dr. Daah's hypnotic trance indicates that in some cases a majority rule basis was used) on the characteristics of the perfect human being.

It is this committee of designer gods that explains why some men want beards but can't grow beards while others always appear unkempt with stubble not only on their faces but all over their bodies. It explains why some people are "bow legged" while others have huge toes. It also explains why all women are born with perfect 36 C breasts and all men have 10" penises. The scrolls tell of one tier 2 Indian god who proposed a Shiva form endowing us humble humans with 6 arms! Clearly four more than the average person needs, thus the Intelligent Designers had to decide how many limbs were ideal. In the end, the committee came together and provided humans with the perfect number of arms and legs and breasts, etc. As a man Dr. Hagen Daahs has questioned why this committee would accept the notion that two female breasts were considered sufficient? "Clearly, for women, wouldn't three or five be better?" he asked. In addition, Dr. Daahs has argued that 36C size breasts, while pleasant to look at, are not nearly as useful as size 42D. "Wouldn't size 42D enable sufficient milk storage for several days? In addition, they could be used for self protection" he noted. Our research indicates that is more likely the gods assumed that since each man was endowed with only two hands he would need only two breasts to fondle simultaneously, one with each hand. On the other hand, Dr. Ima Fükuppe countered this argument nicely with the query as to why women needed more than one breast and why men had nipples at all? "What possible use is the male aureole," she asks. "Did the committee just plan ahead in the case of men's nipples but then simply forget the intended purpose?" she asked. Perhaps, being males, they wanted the same symmetry as was designed into the female body. Dr. Fükuppe has been continuing her analysis of why males have any nipples at all and has concluded that the only rational purpose served is to provide males with something to assuage their egos for being devoid of any useful breast purpose. These are just a few among the many unanswered questions our team is faced with. Dr. Daahs has agreed to undergo still more hypnosis in order to better comprehend the god's thinking process leading to all these mysteries.

This concept of design by committee would have been expected to yield a few variants of the human species. However, it fails to explain why humans are so very different one to each other, thus supporting the basis that evolution actually played a major role. Are the gods themselves so very different one to the other?

CHAPTER 2

· · · · · · · · · · · ·

The nature of godliness or
"What is the optimum persona?"

THIS IS A GOOD POINT of departure for the rest of us mortals. Let's consider the nature of the intelligent designer(s). It is totally rational to assume that there is no single omnipotent god, isn't it? Since one god creating humankind in his image would have had only either a male or a female to image create from? To be sure this God like entity would have to possess supernatural knowledge of all the laws of physics, of genetics, of the universe and, certainly, human nature. If one were to accept that the Intelligent Designer had this supernatural knowledge, then one could postulate that an Intelligent Designer would be capable of optimizing anything he touched. If this postulation were true then one should also ask why did he design his only son as such a "non impressive" entity? Why for instance, didn't he make him a handsome, rugged, all knowing impressive looking male? Why did he choose to make him Jewish? Nothing against Jews in particular, but there are other nationalities that tend to be more attractive, aren't there? I mean wouldn't Mary Magdalene as a blonde Swedish bombshell have been more appropriate? Sort of like Tiger Woods' ex-wife. Why doesn't God look like the Brad Pitt type? If the purpose of humankind is to propagate the species, Brad and Angelina Jolie would have made a great combination trying, but alas failing, to resist the human longing in the Garden of Eden. Instead, the intelligent designer designed a rather non unique person.

Was it his intention to show that mediocrity in human attractiveness was a virtue? Surely this is the anti thesis to evolution which posits that mutual attractiveness was essential to the pairing up of female and male humans who would come together to procreate and therefore keep god's imaged creatures on earth. However, in favor of the intelligent design theory, none of the other deities mentioned previously are particularly attractive either. Although in further favor of the Intelligent Design theorists, in the early days of humankind it probably did not matter much how mutually attractive the males and females were as any port in a storm would do! After all, if Adam and Eve were the only people alive, would it have mattered if Adam was a skinny 4 foot tall person with pimples or if Eve were an ape like female with 4" hair over her entire body except her pubis, or if she had 12" hair on her pubis? Perhaps it was intended that the truly beautiful people should be the misfits of the human race, the less physically attractive among us the more normal the race of humanoids. Perhaps even the mere classification of humans into classes of attractiveness proves that we are the work of an intelligent designer. A designer that intended for us to struggle with mediocrity to realize that mediocre is the repository of the inherent beauty of life itself. Those among this world who are not damned by good looks often seek solace in education or become experts in their chosen professions. They are indeed often the better of the human species. Of course being born beautiful does also require sacrifice on the part of all such humans. Humility has to be mastered unless one uses their good looks to further their position in life. Is it possible that an attractive female posing nude would benefit her aspiring acting career? Probably not! Surely such evil intent was not created as part of the human personality.

Mediocrity works. Think about the situation wherein you are amongst a dozen or so humans on the planet and the urge to procreate (or maybe just practice) comes on. Who among us would not choose the Fred Thompson type over the Brad Pitt type? Who amongst us would not choose the Julia Childs persona over Angelina Jolie? Given a choice everyone would choose the mediocre person to procreate with rather than the "looks good for a day" type. We have been imbued by the Intelligent Designer to realize that good looks fade with time, but virtue and values last forever. Thus all humans always choose the less attractive but more cerebral among us to mate with. By doing so we perpetuate the human species as God intended, with increasingly less attractive offspring leading to less procreation and less humans. This in turn, leads to less population using less of the Earth's precious resources precisely as the Intelligent Designer intended! When a critical mass of just a few humans was left, procreation would take place without any regard to beauty.

To test this theory, and in keeping with scientific practice, Dr. Fükuppe temporarily established a service in which the sexual favors of women could be bargained for by lonely men. To maintain our scientific integrity Dr. Fükuppe hired women at random by perusing a local phonebook. She hired women from the age of 21 to 81. She the advertised this test event on the internet and printed thousands of brochures which were distributed everywhere. The majority of these were distributed on the streets of Las Vegas. Dr. Fükuppe then took mostly male clients and had them view photographs of each potential female sex partner. The intent was to determine if men would choose on the basis of beauty or whether intelligence in design was the dominant factor. The results of this test were quite revealing. Over 74.2% of the male clients chose the younger, thinner, more bosomy sexier escorts every time, indicative of evolutionary choice dominating over intelligent design. Clearly the instinctive drive to procreate was subliminally driving their choices. However, when Dr. Fükuppe also polled the older, wiser escorts, those males over 60 years old, her results indicated that they would be happy with any of 99.9% of all the female clients that were available. Clearly this finding shows that the intelligent designer imbued each human with clarity to choose mediocrity over beauty as males mature thus gaining great wisdom from their age.

Unfortunately for evolution theory, most of these older men were already well beyond their reproductive capabilities thus invalidating the procreation intent. Still, wisdom prevailed. Lesbians preferred women over 99 to 1% while gay men preferred men over women only 76% of the time. This finding surprised Dr. Fükuppe as it provided another strong indication that gay men choosing women over men sometimes are driven by their need to procreate thus supporting evolution over their achieving selfishly pure sexual satisfaction. Of course the predilection for gay men to have their partners' semen deposited either orally or anally distorted the evolutionary theory in favor of intelligent design, which clearly does not favor homosexuality and therefore intended that anal or oral deposits of semen will lead to the definite extinction of homosexuality. It should be noted that thousands of years of homosexual sex has almost completely eliminated homosexual persons in the year 2011. It appears that those who are opposed to gay marriage seem to have missed the point that such couples will, eventually, drive themselves to extinction as they do not reproduce in the manner intended by the designer gods. As a factual matter, they do not reproduce at all if performing only homosexual sex acts. Dr. Daahs has indicated that he has found no evidence of homosexuality resulting in new humans and questions whether homosexuals would have existed a single generation ago, raising the question of whether homosexuality has arisen as a new god creation within the last 40+ years. During one of Dr. Daahs' hypnotic trances, it was revealed to him that homosexuality was in reality a "design option" put into humankind to create fear among the religious followers in order to make their followers more amenable to coercion to the laws of the designer gods. Evolution appears to have circumvented this design intent.

This study also revealed that women preferred "clean masculine looking, non smelly and reasonably endowed" men over "dirty, smelly but well endowed men (penis sizes in excess of 3" circumference and 10" in length)" by over 92.56%, another indication that intelligent design failed. Procreation was not the sole choice driver! With such mixed results, the intrepid researchers continued their search for absolute truth.

Dr. Hagen Daahs poses the question, "if homosexuality is a choice, why haven't they all died off since they are incapable of reproduction?" This question also intrigued Dr. Fükuppe so another experiment was created. If intelligent design is correct, then all men should be interested in performing a sex act solely for procreation! To test this theory, Dr. Fükuppe took pictures of very attractive females, herself among them strictly out of curiosity. She then showed these to gay men while measuring their penile arousal state with instruments. Amazingly, only 4.7563% of the men tested had any response. Clearly species propagation was not the goal of these individuals. Dr. Fükuppe then manipulated these photos using a popular photo editing software program and added male reproductive organs in place of the vaginas which were, of course, prominently displayed on these same women. The gay male response grew to almost 96.3%. When she left the female breasts in the picture, the percentage of men with a penile reaction rose to an astonishing 99.6%! Clearly the lack of female reproductive organs did not deter these individuals all of whom clearly preferred sex to reproduction. Further research indicated that the predilection for the gay lifestyle could result when the DNA from two normal heterosexual persons combined to form a new human. Research has shown that when a female possessing male Yang character trait and a male possessing Yin character trait create a new life, there is a 1:

250 chance that their offspring will be homosexual. This research indicates that marriage between a man and a woman should only be allowed by society if the female is purely submissive and the male is purely dominant, thus ensuring survival of the species without gay persons resulting from the liaison. Many church leaders are considering this possibility and are contemplating a revision to the bible, to be called the "The Gospel According to Brucie." Dr. Fūkuppe has been researching the characteristics that religious leaders could use, pre marriage, to determine if the resulting offspring will be gay or not. In her ground breaking research studies she has determined that a key characteristic necessary for females to avoid homosexual offspring is the ability to tolerate a males desire to party with guys rather than his life's mate; the intensity of his desire for stock car racing; his inability to be faithful to a single female and his desire to drive pickup trucks. The key characteristics for males to possess to avoid becoming extinct is their desire to please their mate with constant loving attention; verbal recognition of the female's dress habits; a strong predilection for cunnilingus; a tendency to shower the female with gifts frequently and the ability to shower at least once per week. Based on these findings, Dr. Fūkuppe believes that the institution of marriage between individuals is not likely to produce LGBT offspring.

What if the intelligent designer was so smart and so brilliant, that he chose to make the son of God an ordinary person? If he did this, then Jesus could identify with humankind and struggle with all the crises that any teenager has. He would struggle with the adrenaline rush of a good game of "hide and seek" or struggle with his testosterone and other hormones when he first saw Mary Magdalene looking like a movie star, of course. Had he seen her as Julia Childs, maybe not so much? He probably suffered from acne, just to teach him how to be humbly human and to spend his teen years wondering if he would ever grow up and what he would be when he matured, i.e. fireman, goat herder, cheese stirrer, Rabbi, Indian Chief, pilot, investment broker. It was very important for him to be human. If he were an obvious god like creature he would have had to answer to all the critics as to why he wasn't making life better and faster; or why he wasn't just ridding the earth of the Romans instead of allowing himself to be crucified by them. Indeed, as a true son of an intelligent designer he would have had to change the world he lived in. Apparently however, the intelligent designer had other plans. The intelligent designer intended that human kind learn slowly, over eons and to suffer and to wonder forever why are we are here. Perhaps the intelligent designer even intended man to learn from bogus theories such as "evolution" or "equality" or perhaps they enjoyed the endless war that humankind has pursued over many millennia. (Note: for most of the world's religions, these wars are only a little over 2000 years old).

A skeptic (not so the Intelligent Designer theorists!) argue that perhaps Jesus is a myth. He is not a creation of the intelligent designer. Instead the entire bible is just a good novel created by a few smart Bedouins who were wondering how to support their families living in the desert with all that wasteland going to waste. Instead, like any good Bedouin, they saw the potential of creating a good story about a not so intelligently designed person fighting off the scourges of the world and turning their wasteland into a tourist attraction. Give them credit, wow did it work! Of course it took a little longer than they expected, but then history has shown us that all small business ventures always take longer and cost more than initially than expected. This is evolution at its very best.

Dr. Hagen Daahs also postulated that perhaps the omnipotent god did intend better results but since man was created in his image and perhaps if God was not infallible, then he simply made a mistake. If man was truly designed intelligently in his image, then it follows that perhaps god is also flawed. If God were not so intelligent, then maybe he just had a bad day, and wanted to appease one of his gambling buddies? You know like, "watch me have some fun with humankind, Thor." After all, why would an intelligent designer not imbue his creation with extraordinary abilities?

Another logical reason for the intelligent designer to not show his true abilities are that maybe he just isn't that smart. Maybe he is very human-like and therefore only possesses average intelligence. Perhaps he is just another average entity with the ability to hire out the intelligent design tasks. Sort of subcontract them out to various god like engineering teams (Heavenly Halliburton?) that may have been assigned the task of designing, say, the not too intelligent cows or the prairie dogs or flies or anteaters, all of which had to be completed within seven days. Another lesser team may have had the task of designing the not so brilliant species known as sloths or even the now extinct dinosaurs. It is perplexing to consider that the omnipotent gods did not foresee the difficulties that they created with the dinosaurs, having designed them to go extinct. Did the intelligent designers also deliberately design mosquitoes to be so annoying and deadly? Does mankind need the mosquito or could we have more butterflies instead? Perhaps the god design team just wanted to get their seven day chore over with so they could go golfing. Perhaps they just needed the seven day design gig for the money to afford their vacations on another galaxy and the intelligent designer just owed them a favor and had the human creature done quickly and on the cheap. Probably not so much!

CHAPTER 3

• • • • • • • • • • • •

Intelligent Design versus Evolution—
or, shit happens

ONE OF THE MORE INTRIGUING aspects of intelligent design involves exploring the design of the human body. After all, aren't we all designed after the perfect image of God? Is it difficult to imagine an omnipotent being designing our bodies to be less than his, less than perfect? Yet, are we less than perfect? Is there a reason the intelligent designer did it that way? The brilliance of the human body design is overwhelming. Do you think the gods needs to eat every day? Do you think the gods need to defecate several times a day? Do you think the god's shit stinks? In this chapter we set out to answer the question, "Where is the brilliance in this design?"

With all due respect, Dr. Daahs posits, "If I were god, I would have designed me a little bit different. Like I would require my body to require only chocolate ice cream in order to exist. As a matter of fact, I would simply redesign my taste buds to only respond with chocolate flavor regardless of what I eat." Of course, Dr. Fükuppe immediately countered with a theory of her own. "Why chocolate, she asked." "If I had a choice, I would have designed males to survive only by licking the protein from between my thighs. Of course", she added, "it could taste like chocolate if that was important to men."

"Why do we only digest a small portion of what we eat," perused Dr. Daahs. If we were designed by a genius, wouldn't we have been designed to process 100% of our intake of food, thereby, producing no waste or fertilizer? Perhaps animals could be designed oppositely by producing 99.9% waste and therefore generating lots of fertilizer. On the other hand perhaps God is brilliant. Therefore he designed us to require a variety of foods. He required us to need animal proteins so that we could learn to subjugate and slaughter helpless animals for food. It is obvious that cattle have no real use on Earth if they were not raised for human food. Of course one could argue that cows are necessary as they provide milk for babies adopted by gays and lesbians who cannot produce breast milk. This idea, however, would suggest that the designers had the LGBT community in mind during their original design phase.

Cows are slow moving. Think about it, have you ever seen a picture of a cow running? They have no aspirations to become house pets and only eat and defecate, polluting the air with greenhouse gases as they perform this act. Clearly, if God had intended that cattle be designed for any purpose other than human food, the animals would have been endowed with horns, swiftness of foot, the ability to jump fences in a single bound and perhaps the ability to change their coloration better to hide from predators. The intertwined need that humans have for animal protein coupled with the presence of animals that appear to have no other purpose in life other than to serve as food would argue the case for the brilliance of intelligent design. People crave meat; cattle exist to satisfy the need for meat; ergo, brilliant design. Well, close enough although, he could have eliminated the fat and gristle from cattle meat as well.

Another argument supporting this theory is that if evolution were a reality, would not the predators like saber tooth tigers have eliminated the species known as cows? Would they not have served as great meals, especially the Prime Ribs for saber tooth tigers? Thus the proponents of evolution have produced their own downfall by being unable to account for the existence of cows on an evolutionary basis. From what species did cows evolve; Fish, sheep, pigs, dinosaurs, or flying dinosaurs? From what species did sheep evolve? What is the missing link in cow evolution and why have we not found the fossil remains of this missing link? Their arguments that man has protected and domesticated cows and therefore enabled the species to prosper does not address the issue of prehistoric precedence which would have eliminated cows long before man domesticated the wild cow herds that dominated the planet. Score another argument for Intelligent Design!

While we were ready to accept the intelligent design win for cows, we were surprised by a report recently received by Dr. Fükuppe from an old associate. A fellow researcher, Dr. Sing Chow Main at the University of Chung Pao in Northern China. Actually Chung Pao was formerly the enlightened nation of Tibet, if you follow international politics. Dr. Chow Main has put forth an evolutionary theory based on old remains that were unearthed in 2008 in Chung Pao province.

According to Dr. Chow Main, he has discovered the fossilized remains of what appears to be a precursor to the lowly cow that appears to be an ancestor evolved from the dog family. The fossilized remains show the canine teeth and strong muscular jawbone features common to all dogs. Astoundingly, these fossils also show that the creature had an udder! The udder was quite large (2.8 liters) with eight nipples likely prevented the canine from running fast or jumping high. Thus it is surmised that the male canine-cow creature with minimal udders, were forced to hunt for food while the female canine-cow was forced to stay around the home, growing fatter and taller over generations and providing canine-cow milk for the babies. It is further surmised by Dr. Chow Main, that these canine-cows were not very agile and therefore not likely the greatest of hunters. Consequently, the females gradually evolved from being meat eaters to being grazers of common grasses in order that they provide sustenance for their small herds of babies. The canine teeth would have been a rapid evolutionary loss changing from canines to molars enabling the canine-cows to cut and chew the grasses. Further proof of this evolutionary step is the determination showing that the female canine-cows greatly outnumbered the male of this particular species. Perhaps the males were only kept as a means for ensuring a sperm donor was available to this herd as they evolved from hunters to be grazers only. From this primitive canine-cow link, came the ultimate evolution into a fatter version of the canine-cow later domesticated by the Neanderthals as a source of protein that endures to this day. However, there are many scientific scholars who doubt this theory, despite the fossil evidence, and because of this controversy we are not suggesting that evolution and not intelligent design is the ultimate source of the cow. Dr. Fükuppe visited Dr. Chow Main in Kung Pao province and reports that the remains of the canine-cows are extremely well preserved and that as further excavations continue we may actually find an entire herd from which further scientific truths can be elucidated. If this theory proves correct, then it may be possible to postulate that many animals evolved in a similar manner; perhaps sheep, for instance. What if they also evolved from human shepherds mating with canine-cows for instance? Over millennia, the sheep evolved uteri that adapted to the larger human penis instead of the penis associated with sheep dogs, thus accounting for another evolutionary step in the animal world.

"If God was the intelligent designer," posits Dr. Fükuppe, "why did he design us with superfluous organs?" It appears that the medical profession has been acting in a very non god like manner by performing useless surgeries to remove body organs that God thought were essential. "Would an intelligent God have put all of those extra organs in our body to cause medical problems?" Dr. Fükuppe queried. "Like, why give us tonsils or adenoids? They are removed at the slightest hint of inflammation and apparently people live a long time without them. Why did God give us appendices? I had mine out when I was only 21 years old and don't miss it one bit," she noted. And then there are gall bladders which apparently we can live without forever. And men can have their sperm ducts tied off and sections removed completely without any ill effects! We have often wondered what happens to all of the millions of sperm that can swim in the testicles but can't go anywhere. It is like living your entire life in a hot tub installed in a dark room. Or women can have their ovaries removed. These are good removals, of course, as they allow for all mannerism of sex acts without any repercussions at all. Dr. Fükuppe reports. "I had my ovaries removed along with my appendix and can have sex with any person I choose at any time! Everyone should have them removed as early as possible," she noted. All of these useless organs. Perhaps they were valuable on another planet in another lifetime before god created humans? "I wonder", muses Dr. Daahs, "why we were created with teeth that can be infected so easily and are so very painful. Couldn't we have been designed with metal teeth that last forever? I wonder why I have to contend with gingivitis as well? Surely these are the result of evolution and not design. Poorly evolved evolution."

Dr. Fükuppe questions the intelligence of requiring two male testes and only a single penis? She also questions why males have any breasts and nipples at all since they apparently serve no useful purpose. Even evolution seems to fail the reasonableness test as two testes and male breasts can only be explained as required symmetry during genome growth. Dr. Daahs asked why if symmetry were the driving force, why don't females have two vaginas and males two penises?

"It would make for far more interesting sexual encounters thus ensuring species survival" he noted.

One of the other thoughts that comes to mind is why did the intelligent design architect put the organs that we use to dispose of waste products so close together? Perhaps this is the result of economy of body area; after all shouldn't the perfect human form be of the smallest size possible. Dr. Fükuppe is quick to point out that the male penis and female breasts should be exceptions to this rule. Perhaps the body should have been the largest size possible, like larger than man eating predators? If man were designed intelligently as being forty feet tall, then we would not have been prey for so many larger and less intelligent animals. What is the perfect size? Are 4'0" tall people ideal? They do not waste as much body heat, have fewer tendencies to break bones when they fall down, and require less food, less habitat protection and do not pass on as much dung. Why then do humans grow taller with every generation of evolutionary impact? Being taller enables us to see over higher obstacles without climbing, we can run faster to escape predators, generally Americans can see over the heads of Japanese persons easier, we can excel at basketball and we can bump our heads on more overhangs. Dr. Fükuppe has performed another interesting experiment wherein she recruited men and women of various heights and had members of the opposite sex rate each individual as to the "desirability" of their bodies.

For this experiment, women less than 4'6" tall were allowed to walk provocatively down a runway completely nude while men rated their appearance and desirability. The overall desirability rating achieved was 86%. When she replaced the women of 4'6" stature with women over 6" tall, the desirability rating was an astounding 99%! Yet taller women were not at an advantage from the point of view of intelligent design. This intrigued Dr. Fükuppe so much that she designed another of her famous experiments to elicit the absolute truth. In her study, she found that many scholars, including religious ones, believe that the human body, like so many beautiful objects, was designed around the "divine" mathematical ratio of "Phi" or that they possess a ratio of 1: 1.62. This magical value can be described in its most elemental understanding if the reader studies the graphics in the figure shown below:

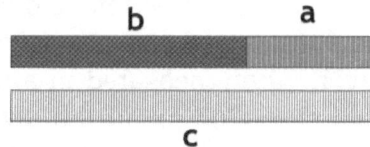

The "Divine Ratio": b is to c as a is to b

We digress a bit here. If a person measures the total length of line "b" in the figure it is equal to 1.618033988749895 . . . the length of line "a." Line "c", in turn, is 1.618 . . . times the length of line "b." Such precise mathematical terms can only be of a creationist basis as early man would have rounded it off to approximately 1.5:1, lacking precise measurement devices for a very long time. This ratio occurs frequently in natural objects and certainly is considered the most attractive ratio used in art. Thus 1.62 is regarded by many persons as "divine" or "golden". God created! This is the ratio found in the dimensions of the Greek temples such as the Parthenon and Roman structures as well. It is this ratio of width to height that imbued them with symmetrical beauty. Some believe that DaVinci has incorporated the divine ratio in the certain of his paintings including the Mona Lisa. In the human body this ratio can be found in the measured distance between the top of the human head to the bottom of the fingertips. Think of this as line "a." Then line "b" is the measured distance from the top of the head to the navel! Do you, dear reader, fail to see a pattern of divinity here?

We could not help but wonder if the divine ratio was a creation of the omnipresent God, then it would be everywhere. What if the remainder of the human body was designed with other such ratios? Drs. Fükuppe and Daahs immediately undertook an extensive study to determine whether the divine ratios of 1.62 existed throughout the human form. This would be a strong argument in favor of Intelligent Design. Dr. Fükuppe immediately travelled to a neighboring Colorado college and recruited over 1400 students for her study. She undertook an extensive effort to measure portions of their body to see if the divine ratio could be found. Dr. Fükuppe insisted on handling the males while Dr. Daahs assisted by measuring the females so each could focus on this laborious and difficult task.

Dr. Fükuppe initiated her scientific experiment by measuring the distance from the male ass to the front of the flaccid penis and compared this distance to the distance from the male ass to the navel. Alas the golden ratio was not present. She measured the distance from the top of the male penis to the bottom and compared this to the related length of the male testes and alas also failed to find divinity in this relationship. However, when measured in a hot rather than cold temperature environment she often found the divine ratio to exist in these relationships. She then saw the fallacy of her test and performed another set of measurements. She measured the length of tumescent penises and compared them to the distance from the base of the erect penis to the outermost distance of the ass. Amazingly the results indicated a strong correlation with the divine ratio of 1.62. Dr. Daahs then measured the distance from the tip of erect breast nipples to the chest and compared this to the total frontal body width of the female subjects. In addition, being a scientist, he measured the protruding distance of the female nipples from the breastplate and compared this to the thickness of the body from the base of the breast to the backside of the person. Incredibly, both of these measurements were closely correlated to the divine ratio of 1.62! Dr. Daahs is pursuing this investigation by measuring female bodies both during and post menstruation as well as extending the data base to include post menopausal women. To date, the results are nothing short of divine. The correlation of divine ratio to actual body measurements is in excess of 98%, a pseudo scientific fact of enormous impact.

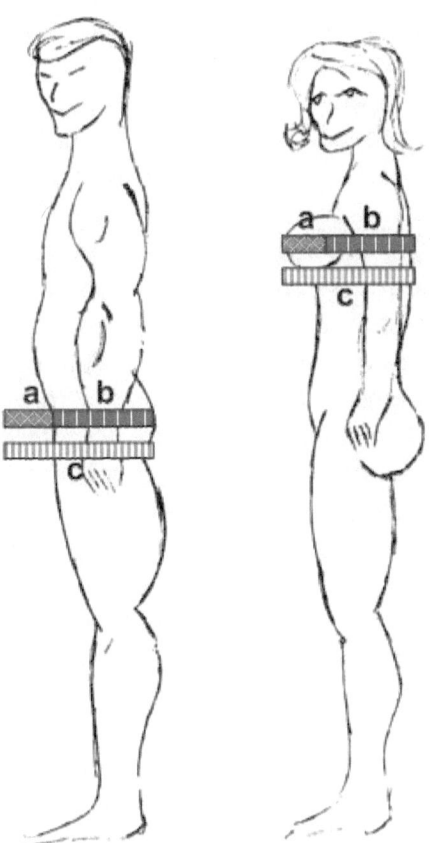

These results were so astounding that the intrepid researchers felt compelled to verify them with an additional data base of over 7200 students from 4 colleges and universities. "I love conducting pseudo scientific research" Dr. Fükuppe noted. The scientific as well as the amateur community has also jumped aboard this phenomenon and these tests are being replicated across the world on a global basis as this book goes to press.

Both researchers then posed still more questions, such as "are breasts designed to be in proportion to the overall height?" More critical to Dr. Fükuppe's analysis was the question "are penises designed in direct proportion to height?" she questioned. Thus was born still another of her experiments. The results confirmed Dr. Fükuppe's hypothesis. Women with heights of 4'6" or less had a preponderance of size 34B breasts. Their vaginal depth averaged 12.5 cm. On the other hand, women over 6' tall had an average breast size of 36C and vaginal cavities of 14.6 cm. Short men (under 4'6") tall had average penis sizes of 15 cm in length while the taller men averaged 16.7 cm. Unfortunately these results failed to confirm the Intelligent Design theory since male and females of the same approximate size seemed to have correlated sex organ sizes. On the other hand, over 98.3% of all under 4'6" females chose men with larger penises despite their mismatch in sex organ correlation. Males had a similar result. In over 89.9% of the cases 4'6" males chose the taller females with larger breasts and deeper vaginas in spite of the fact that they could neither place a tall woman's breasts in their mouth during intercourse nor could they satisfy the female desire for "fullness" during intercourse. Dr. Fükuppe further noted that the sensation of "fullness' was a major factor in her personal conduct of this testing which resulted in a desire to complete the procreative rituals with the largest of male participants. The test results show that the intelligent designer intended perfection in mating should result in pairing of individuals of the same stature. Evolution, on the other hand, indicates that the species is still growing taller in pursuit of being able to run ever faster.

CHAPTER 4

• • • • • • • • • • • •

The body perfect or "I would have designed me differently"

ONE OF THE MANY SILLY design features we humans have is that we were designed with an interconnected hole in our head through which we breathe, eat, taste and smell. How novel. If you happen to swallow something too big to get past your air passage, you will choke and die in less than 120 seconds. The intelligent designer took care of this possibility when he created a Dr. Heimlich who has taught all of us to perform the "Heimlich" maneuver" to dislodge such items from the air passages and therefore reduce the threat of death by choking from our lives. Unfortunately a great many pre historic people died from choking before Dr. Heimlich was able to teach his theory principally because evolution is such a slow process. It would have been far more intelligent to have taught this technique to Adam in the garden in which case evolution may have passed it on. On further thought it would have been far more beneficial to have taught it to Eve as men tend to eat much too fast and women are better learners.

Using one organ for multi tasking fits well into our 21st century lifestyle. Never the less wouldn't it have better to let air go directly into our lungs rather than pass thru the throat? Much more importantly, wouldn't life be more pleasant if the sense of smell could be turned on and off at will? "On" for flowers, "off" for public toilets or "on" for roasting chestnuts and "off" for flatulence!" Is it not also brilliant that when a human has an illness producing mucous in the nasal passages, it conveniently drips down directly into this single orifice, the mouth or the throat. The brilliance of this location is noted when one considers what would happen if the nasal organ, otherwise referred to herein as "the nose," were located on the back of our heads. It would then drip mucous down our backs where it could not conveniently be consumed as it can be with the mouth allocated directly below the nose. What if the Intelligent Designer would have foreseen the inevitable effect of the nasal passages being clogged when we sleep? Could he have eliminated snoring? What if he simply put the nose in charge of breath and the mouth directly connected to the stomach so that the two never linked together? This would have the secondary effect of keeping vomit in the mouth only and not having it stick to the mucous passages of the nose; Evolution or intelligence in design?

One of the other thoughts that comes to mind is why did the intelligent design architect put the organs that we use to dispose of waste products so close together? Perhaps there was a practical reason for the placement of these organs. This seems like the most plausible explanation since animals are able to dispose of bodily wastes from the standing position with both liquids and solids simply coming out and falling to the ground. Perhaps these waste products can even be eliminated while walking or, as with birds, while flying! It should be noted that Dr. Daahs has confirmed that the observance of the act of birds defecating while flying was the precedent from which mankind invented aircraft and bombs. We wanted to verify if humans could perform like animals thus providing our research with another basis for intelligent design theory since evolution cannot account for this phenomena. The scientific quest for knowledge overwhelmed us and we designed a crucial experiment. Consequently Dr. Fükuppe and I performed such an experiment wherein we ate and drank constantly for days while we walked for 120 miles, nonstop. Our theory proved to be precisely correct in that we did not have to stop except for very brief periods from our walking to rid ourselves of our bodily waste. Instead we were able to walk for several days, pause slightly, eliminate wastes and continue. However, occasionally, we were forced to stop to cleanse ourselves of a horrid, clinging, rigid mess that was sticking to our bodies. Dr. Fükuppe, in particular, suffered as her boots were constantly filled with urine and feces after only two days of hiking. Perhaps we should have accepted the realization that we humans might have been designed less perfectly than animals and therefore are required to at least stop and stoop to avoid making a mess over our bodies.

"It is unfortunate," Dr. Daahs posits, "that we were not designed with an anal opening just above the right heel and a similar urinary opening on the left heel that would enable waste disposal without all of the fuss and inconvenience created by crouching" he noted. "In addition, we could wash ourselves very conveniently by placing our feet into a bucket of water rather than having to bathe our entire bodies," said Dr. Fükuppe. On the other hand, the waste tubes would have to have been designed to travel down our legs thus slowing our ability to run. Joggers would have a terrible time exercising and instead of jogging they would have to practice walking on their hands instead. Putting the waste tubes down human legs would also have necessitated another location for the sex organs. Redundancy of purpose and the resulting efficiency of design would have been lost; not very intelligent! However, in our search for reasoning, it appears that there is an explanation worth considering. After all, the proponents of Intelligent Design can't be totally stupid, can they?

Clearly a rationale must exist. If we consider one of the most basic tenets of intelligent design, the implication that the Intelligent Designer acted to produce rational and religiously righteous human beings, then perhaps the reason for the placement of such organs has a sexual reason for being; sure enough. After much thought it is clear that the intelligent designer intended that humans have sexual intercourse only for the selfless reason of procreation. Clearly he (they, she?) intended that we only have sexual intercourse in the "missionary position" with the man on top and the female as a submissive, or non-submissive, partner in the act itself. The reader should note that the missionary position is the dominant male position of control as every alternate Kama Sutra position requires some degree of female acquiescence, a pre-requisite missionaries could not always achieve! However, Dr. Daahs notes "not all missionaries practiced sex in the missionary position." Many preferred the rear entry position with the female bending over at the waist or doggie style. These soon became known as the 'canine cow missionary prayer' position." Dr. Fükuppe queried as to why "missionaries" were concerned with sexual practices at all, after all weren't they practicing god's teachings?

Surely the designer would only have included procreation as this motivation and eliminated all other sex acts as being deviant. The designer may even have suspected that evolutionary change might impart to humankind the desire to exploit sex for purposes beyond procreation, even to the point that humans might engage in sexual practices for pure physical and psychological pleasure. One of these pleasures is the deviant behavior between two persons, assumed to be of opposite sex but homosexuality is itself still another consideration, wherein mutual oral sex is performed in the infamous 69 position. Knowing that such deviant behavior might have resulted from the effects of evolution, the designer deliberately placed the anus within inches of the female vagina, thereby forcing man to perform such deviant acts as oral stimulation with his olfactory organ within mere inches of the female anus. While modern soap and hot showers ameliorated many of the unfortunate happenstances that can occur while performing cunnilingus in this position, the cold mountain streams of millennia past would have rendered this act less than satisfactory on many, many occasions. Clearly this organ proximity would discourage the lascivious act from being performed! Hence the mystery is solved. The Intelligent Designer was prescient enough to anticipate the effects of evolution and to place impediments in the path of sexual deviation. The placement of the sex organs, and indeed use of the sex organs as waste disposal organs has virtually eliminated all vestiges of oral contact to these organs.

This pseudo fact was confirmed in a further recent study conducted by Dr. Fükuppe and the University of Baccala, which showed oral sex acts were only performed under duress situations in our modern 21st century society. This study was exhaustively conducted over 2.5 hours with employees from a barn cleaning crew and sewage treatment plant at a giant hog slaughtering factory located in Southern Bulgurossa. As many as 45,000 hogs live in this facility and over 500 can be slaughtered every day. As the men and women completed their work duties, they were solicited and financially incentivized to perform oral sex upon their fellow workers. 98.8% refused. Many remarked that it would have been a far more intelligent request if they had been able to turn the olfactory sense "off." Despite the incentives of $10.00 dollars, oral sex was clearly not a condition that evolution had modified so as to render it "desirable". Other studies conducted by Dr. Fükuppe have indicated that latter day evolutionary impacts on college students indicates that oral sex may be undergoing a transformation and that it may become slightly more acceptable than indicated by the University of Baccula study.

Dr. Ima Fükuppe has conducted an extensive internet search and analysis around the world that further confirms the apparent non existence of the oral sex practice, except on isolated North American college campuses. Dr. Fükuppe has also personally participated in a personal research project in which she encouraged various men to permit her to perform this deviant act and selected scientifically interested friends to perform oral sex upon. She has recorded her results and provided her analysis to the general body of scientific knowledge. The results of her testing confirm the factual existence of an Intelligent Design. Of 1000 men solicited by Dr. Fükuppe, only 920 agreed to let her and her team perform the act, thus confirming that it is not an inherent or dominant trait in the male genes. The results of this test clearly indicated that the 80 men who refused Dr. Fükuppe's generous offer did so because they were "married." It is postulated that these men only participate in procreative missionary position intercourse sex and do not indulge in sex acts only for personal pleasure. This is in full accordance with Intelligent Design theory since oral sex is an anathema to the intelligent designer as clearly it has no reproductive function.

Another human attribute in which humans have interacted to refine the intelligent designer's intent is in the seldom researched area of male smegma. As is well known, the male body was designed with a penis from which man can both eliminate bodily fluids as well as use this instrument to penetrate a female vagina and deposit sperm from which more of the intelligent designers "look a likes" could be spawned. The penis was a topic of interest to Dr. Fükuppe and so another series of experiments were established to verify whether it was designed intelligently or whether it just evolved in an odd fashion from an early appendage.

Dr. Fükuppe, being of Swedish/German heritage, was in her early research years truly astounded when she discovered that some men did not have a foreskin on their penises. "Mein fatehr circumcised nicht!" she indicated. Clearly the Intelligent Designer designed the penis to have a foreskin. Every male is born with one. Many men die retaining their foreskin. It is yet another body part that seems to perform a function not necessary for life itself. Yet, it was common in some societies to punish their male offspring by having the thin skin that normally surrounds the head of the penis removed by surgery, a practice akin to slicing cold cuts very thinly but totally dependent on the surgeon's skills! Why had the intelligent designer put the foreskin on the penis to begin with if it is not necessary? Was this a case of evolution or not so intelligent design? Fortunately Dr. Fükuppe has personally examined over 1600 penises and has established a data base that can shed light on this characteristic. In her examination of 1600 penises, Dr. Fükuppe observed the following data:

- All male penises can become tumescent during the oral stimulation resulting from such an experiment.
- 0.06% of all circumcised penises experienced discomfort from oral sex.
- Males with non circumcised penises achieved climax within 7.5 minutes of lips to penis.
- Males with circumcised penises achieved climax within 9.5 minutes of lips to penis.
- 73% of men examined were circumcised while 42% were not.
- Women seldom worried about female smegma.
- 100% of men participating indicated that they would volunteer for additional research of this type.

In analyzing this data, it appears that the foreskin is a "good thing" in that the men with uncircumcised penises achieve orgasm faster than uncircumcised men. This would indicate that the foreskin serves the purpose it was intended for in making the head of the penis more sensitive thus increasing the likelihood that male penetration of the female vagina would result in sperm deposition faster, thus not relying on pleasure to prolong intercourse. Unfortunately, evolution appears to have countered this reproductive drive with a predilection to oral sex, which seldom results in species propagation. Clearly the Intelligent Designers knew that the foreskin would benefit males in oral sex situations. However, if oral sex was an unintended consequence of sexual interaction, why didn't the intelligent designer just leave the foreskin off?

What Dr. Fükuppe has found, is that the natural penis, when left unwashed, will produce a waxy substance under the foreskin known in scientific terms as smegma. Note: This substance should not be confused with "Smeagle," which is a character in another book by a lesser known author. Smegma is produced in both male and female genitalia. In males it is composed of skin cells and lubricant and is thought to provide lubrication for intercourse. However, when smegma is allowed to accumulate under the not needed foreskin, it can develop a rather heady odor and further has been reported by Dr. Fükuppe to "taste unpleasant. In circumcised males smegma cannot accumulate as easily. It is therefore postulated that the intelligent designers intended the male penis to have a foreskin. More sensitivity and a clear discouragement to engage in oral sex, precisely as intended. Evolution, on the other hand, would only have produced a foreskin to protect the penis and to provide males with a disproportionate pleasure in sex.

In other instances the designers clearly anticipated the potential for evolutionary type acts to further despoil the beauty of perfect design. By designing the anus to be strictly a waste disposal organ with no natural lubrication, the designers ensured that the organ would not become useful for deviant sexual practices. This was very important as the Intelligent Designer placed the anal opening very close to the vaginal opening. It is natural therefore that occasionally the passion of the sex act would occasionally, only accidentally of course, result in the male penis being inserted into the anus of another human by mistake. The intelligence of the human body design is enhanced, of course, by this act resulting in occasional pain. Clearly the foresight of the Intelligent Designer is critical in that this pain ensured that anal sex would never be performed by humans. However, the intelligent designers, perhaps accidentally, also imbued the human mind with the power of creativity and invention. A power so often abused by man. One of the greatest inventions of man was the water soluble family of body lubricants. It is with the advent of unnatural man made lubricants that the anal organ has been the subject of significant exploration and deviant practices not intended by the original designers. This is particularly true of the Gay community where the presence of lubricants has significantly enhanced sexual pleasures. Once again, evolution is shown to be a deviant force that, if eliminated, would enable natural perfection to exist in its most pure, and holy, form. Perhaps the devil made it so?

CHAPTER 5

. . . .

Sleep, or "why am I wasting 8 hours a day?"

IN ANALYZING THE HUMAN BODY characteristics, another bodily function comes to mind that seems to beg the question, what was he (they, she?) thinking? Have you noticed that you typically spend 6-8 hours a day sleeping? Seems like this time could be better spent doing productive things like having sex or working at another hobby. Why then do we require sleep?

From an evolutionary standpoint there would appear several reasons for this phenomenon. For instance, in the world preceding electricity, it was difficult to see well enough to perform the necessary tasks such as hunting or cave wall art generation. Those pesky wood fires were insufficient and unreliable for all manners of tasking. The second problem faced by early man was the necessity to protect oneself from their nocturnal predators who really were better equipped by evolution (design?) to see and track in the dead of the nighttime darkness. This may be another case in which the intelligent design theorists have a major issue. For instance, why wasn't man created with phenomenal dark vision to detect predators or to stay awake all night? Lacking this attribute, early man deemed it prudent to seek shelter that could be protected and warmed. He also found that opposite sex companionship assisted in relaxation and preparation for sleep through the dark period of each day. He may have also found that same sex companionship provided equally satisfying relaxation, especially on long hunting trips. Unfortunately, pseudo data is lacking to document the bonding practices on long, dark, cold and dreary hunting forays.

On the other hand certainly an intelligent designer would have programmed sleep for the good of his most favored creation. One possible reason for requiring sleep is that the human mind requires the sleep period to spend in the realm of unconscious thinking. Early man certainly required the development and exploitation of extraordinary sensing, the so called 6th sense, to survive. As part of this additional sensing, the need to allow the subconscious mind to sort out the details of the days learning experiences and to prepare for survival for the following day clearly had a requirement that man should lose his conscious state of mind and permit the subconscious to take over. The only effective time of day for this to occur was, of course, the darkness of night. Hence the intelligent designer was way ahead of evolution by creating and encouraging the development of the subconscious mind, a state of mind understood and appreciated by all mankind to this very day.

Another rationale that the Intelligent Design crowd recognizes is the simple fact that the intelligent designer did not intend that humankind live a frivolous lifestyle in which he only worked for 1-6 hours a day. Surely the intelligent designer intended that if man could satisfy his needs for food and shelter in less time, then he would certainly waste the additional hours in inventing games or indulging in other unwholesome activities. Therefore intelligent design had as the intent of the creator to ensure that all of man's daylight activities were restricted to either survival or adoration of the great intelligent designer. This theory has gone somewhat astray in modern times in that survival activities do not consume as much time and humankind is free to spend the hours of darkness in pursuit and adoration of the great intelligent designer. This has led to the creation of institutions and individuals that have been able to garner great wealth from those that adulate for the intelligent designer. These few people now have even greater time to spend in adulation and some are now capable of staying awake 24 hours per day in counting the proceeds of adulation, as the intelligent designer intended. Surely evolution would not have enabled the invention of electricity and the subsequent ability to diminish the role of the sub conscious mind.

Of course it is certainly possible that evolution could have played a part in developing the unconscious mind as man naturally went into a relaxed state at sundown and the brain decided that it could use this time to repair and refresh bodily functions as were depleted during the mind's conscious period. Once sleep had evolved, it was an easy step to utilize this uninterrupted period of time to also develop the unconscious mind.

In order to test this theory of what principally was being developed during sleep, Dr. Fükuppe consolidated yet another test. She postulated that sleep was the time period that the unconscious mind worked through difficult problems and also formulated plans for the conscious state of mind. Intelligent design predicts that the mind will spend its unconscious state preparing for godly worship and for the solving the problems that the person went to bed with. Fortunately a team of researchers at the Institute for Brain Functionality in Zurich Switzerland has been conducting MRI scans of the human brain while subjects are asleep. By correlating the areas of the brain that are energized by certain stimuli, the team of researchers can determine what the subject's mind is thinking about at any given time. Dr. Fükuppe flew to the institute to participate in the groundbreaking research. The research results, as of yet unpublished, indicated that the male brain spends 25% of its sleep time unconsciously concentrating on sports, cars and flatulence. As a matter of factual evidence, these subjects passed gas during their sleep whenever the cerebellum is active for more than 1-2 minutes per hour. Dr. Fükuppe convinced this team to also instrument the male penis to see if a correlation to brain functioning during sleep could be found. Indeed, the results indicated that whenever the frontal lobe was active the male penis became erect and involuntary stroking of the male sex organs took place. This action accounted for the remaining 75% of male subconscious brain activity. Dr. Fükuppe is continuing her work to determine precisely what thoughts are being processed during this time period.

Contrary to this work, the researchers investigated female brain unconscious activity as well. In this state the female subjects had their brains continuously scanned with an MRI while asleep. Their response was considerably different than the males. In early sleep onset the temporal lobes were dominate with the subjects often smiling when music and flower odors were introduced into the test regimen. Brain scans indicated that clothing choices and hairstyles were often studied as the subjects entered a deeper state of sleep. Under no circumstances was the cerebellum active at all and the female vagina never became moist nor was it touched during sleep. However, during female menopause the cerebellum became very active with images of males being tied in dungeons and being tortured were present. Indeed, images were recorded in which the females envisioned themselves as Ninja warriors and often fought with their co workers and bosses over trite details of corporate life such as how hot the coffee is when served and other such matters. Still, the cerebellum controlling sexual urges was strangely dormant. This confirmed the findings of Dr. Daahs when he was placed under hypnosis and determined that the female gods were responsible for the final design of the human mind and they definitely were not interested in female sleep sex.

CHAPTER 6

· · · · · · · · · · · ·

The Incorporation of the stupid gene into the human species, or "The development of Red vs. Blue states"

ONE OF THE TENETS OF evolution posits that all species undergo a trial by enduring life itself. During these trials the travails of daily survival will ensure that species develop attributes that will enable them to out skill their opponents and therefore survive a kill off of the weaker species. This theory is often referred to, laughably, as the theory of "survival of the fittest." This theory posits that if two or more species cohabit within a common space, then they will compete for the basics of food and shelter and procreation. In such a theory the "fitter" or better adapted animal will either out muscle a weaker opponent or, under intelligent design, will have been endowed with god's creative brain so that they out think an opponent. This is evident in today's surviving animal species for instance in the mountain lions who have evolved over time, or been designed from the start, to become stealthy nocturnal hunters. They have superior night vision, roam over a large geographic area to have more prey available, and they have been known to prey upon less intelligent species such as deer, elk, occasionally non evolved humans and other wild animals.

The same evolutionary aspects can be observed in the human species. In this case, evolutionary dominance has both a mental as well as a physical dimensional aspect to it. Why then do we have so much stupidity in humans? The answer lies in an evolutionary trait that was designed into humans by the Intelligent Designers. The answer is DNA, the very blueprint from which the cells create our bodies and mind. DNA creates the very god like beings we are. It determines how long our noses are, how large females breasts become before surgery and how "smart" our brains are. The genes that create all this are slightly modified by the epigenes, which can modify the genetic code during our most formative years, such as ages 1–3. The DNA and the proteins that could determine the resulting intelligence are shown, as example, in the following figure. This resulting genetic code determines our personality as well as our intelligence level.

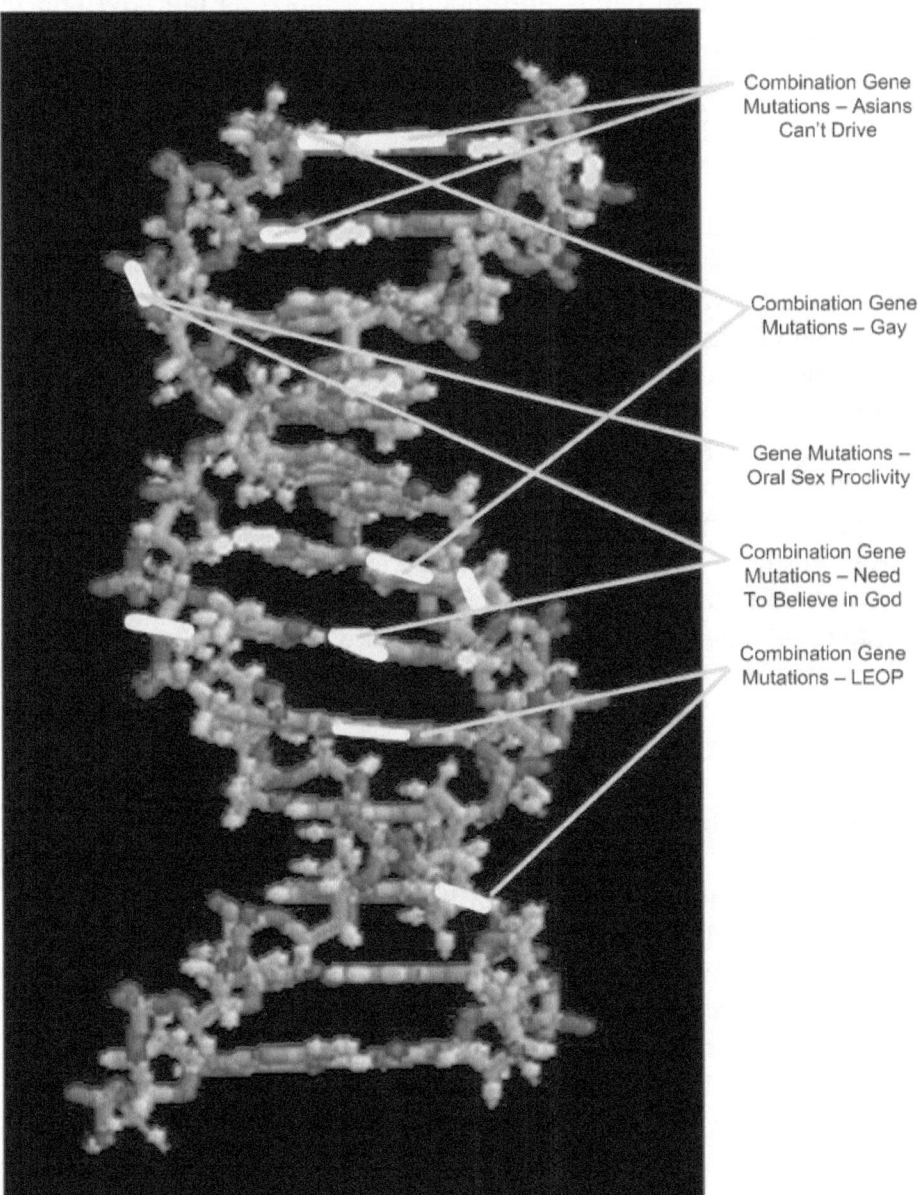

As shown in the figure, proteins (combinations of the basic amino acids that make DNA) exist for all humankind proclivities including dullness of thought, sex, need for a herd mentality, sex deviations, hatred of any human that looks different, love of guns, masculine dominance over dumb animals, masculine dominance over females regardless of species, desire to never travel, desire to never experience new novel thoughts, etc. these genes are known to science as the Low End of The Pool (LEOP) syndrome and are the principal genes that create the mind of the ignorant persons. When a man and a woman possessing LEOP genes mate, these genes combine to form even more powerful gene pools of LEOP and create ever more intense levels of stupidity. The proponents of Intelligent Design believe that this LEOP gene set is good for humankind as it enables the religious sects to lead their followers wherever they choose to lead them. The evolutionists have a difficult time with the LEOP gene set because according to evolution, these people should have been eaten by predators long ago and they should no longer be part of the evolutionary evolution. Dr. Daahs has researched this discrepancy in evolutionary theory. His findings may shed light as to how this was allowed to happen.

What he found was that the early LEOP humans congregated around a farming existence in isolated areas of the country where other humans preferred not to live. Examples of these locations can be found in the Midwest and southern portions of the USA and Afghanistan in Asia. These LEOP types were often isolated. In the absence of many diverse, non related humans, they began to have sex with close family relatives, thus ensuring that the LEOP gene set would not only be passed on but in fact would grow ever stronger with each generation. Dr. Daahs found this trait to be particularly strong in areas of the United States, which tend to be isolated. These areas are strongly correlated with those for whom a strong belief in Christianity overrides all intellectual and independent thinking. The LEOP gene set is much less apparent in areas of the world in which a constant mix of diverse proteins results in a gene pool lacking in most of the LEOP set as it gets diluted with each generation instead of enhanced.

Intelligence can be measured in many ways, but for this book, it is defined as the ability to think critically about life and to not follow the herd's simplistic practices and simply follow the leader over a cliff. Once this simple idea is understood, one can track the level of intelligence by its decoupling to most theological teaching. Clearly, unless a person is devoid of most of the mutated genes, intelligent thinking can be very cumbersome. It is interesting that gene mutation seems to occur as an evolutionary process. If it were a "design feature" would such diversity have been part of the fundamental DNA blueprint? Can a person's persona and belief system be changed if in fact, evolutionary changes to our DNA drive the persona?

CHAPTER 7

• • • • • • • • • • • •

Sex or "sex"

SEX; THE WORD ITSELF INVOKES anticipation of exotica and titillation. It is the subject that can separate the intelligently designed folks from the rest of an enlightened society. As the basis for all procreation, it certainly deserves reverence; as the purist source of physical pleasures it certainly deserves and has our interest! Surely this all important act was at the centerpiece of the Intelligent Designer's thought processes to ensure that God's likeness was to be preserved over the millennia. At the Round Table of the gods then, were gathered all of the great minds of goddom. Perhaps the discussion went like so.

"What shall we do for the human species to be able to procreate?" asked CEO god, Benjamin. "Why do they even have to procreate? After all they will only make more of the same" responded the Ethics Advisor god. "We could just create a few 100,000 and leave well enough alone." "No fun in just creating a successful species," said CEO god. "We need to experiment a little to see what we gods could become" he said. "Should we make them like our Intelligent Designer wants?" Why can't we deviate slightly and have some fun?" was the first question from Sir British god.

It should be noted that in the "Lost Scrolls of Creation" the aliens indicated that they created both male and female gods. However, apparently the design process was so noisy and cumbersome that the male gods got into a macho "I can do it better than you" attitude and the female gods (goddesses) decided to leave the enclave and tend to more important stuff like getting a full body tan with which they could get whatever they wanted from the male gods anyhow. Hence the female body was designed exclusively by males which explain the large breasts and full asses.

"Sex is great but certainly humans don't need all of that pleasure even though I, as a god, particularly like the oral versions. Can't humans just exchange cells orally and replicate" asked Japanese Production god. "Or even more simply, can't we just design them to simply have their cells divide over and over to replicate, like bacteria" was an early suggestion. "They could just divide in two parts, each could grow and prosper and then automatically split into two parts again for ever with no need to do anything or to take any actions." said Assistant god Born Again! "That won't work. How would they keep from over running the meager reserves this planet has then? The earth will be filled with identical billions of creatures" asked demi god Swedish Lemming. "Indeed," said the Intelligent Designer omnipotent god. "But I want them to be perfectly designed," he said. "We are after all, perfect, aren't we?" An immediate objection from Japanese Production god indicated that perfection in gods was not necessarily good for humankind. "Let's show the other gods just how creative we can be and make the human procreation act, well sort of a punishment affair," he said.

In our research, Dr. Fükuppe and I could not postulate a single reason for any species to actually require an act of sex for procreation. In a survey conducted by Dr. Daahs the results indicated that only .011% of all participants actually intended to produce offspring by performing the sex act. As Production god questioned, a simple cellular reproduction technique would work just fine for propagating the human species. It would, taken as a factual event, been a very intelligent design. Perhaps the female gods would have chosen this technique. However, if a sex act were not required then how did it become an evolutionary happening? How did the first sex participants realize that it might be pleasurable? The only viable reason to imbue any species with natural selectivity is to create a better, superior offspring. If the gods created perfection, what more could they possibly have hoped for by way of natural selection?

In terms of spiritual maturity, the human sex drive is so convoluted and complex that only a small fraction of humans can understand it. This remains one of the greatest mysteries of all time. Having decided to have humans employ a sexual act to effect procreation, the gods were faced with the prospect of what, why and how?

To secure the future of the species the gods conceived that humankind should have a ritualistic event to ensure that more humans could be created when needed. It was also considered vital that both the male and female of the species participate in this procreative act less one of the two sexes eliminate the need for the other. If a man and a woman were to cooperate in this ritual then the man would need a specific "something" so that he could get his reproductive "something", thumb, toe, nose, etc., deliberately engaged with a female's reproductive "something" else organ. The ritual had to feel good in the beginning so humans would use the system, but it couldn't feel so good that the only thing humans would do is "toucheachothertocreateoohandaahsstuffguttnicht" said the old German god. So the first system analysis conducted by German Engineer god, was to determine a set of feelings that would make men want to procreate. Having a one-sided male dominated system however might not work as men would always only want sex. Engineer god thought, in good keeping with Yin and Yang principles, that he could balance the male impulses by assigning the job of resisting the male impulse to woman. The gods arranged the DNA molecule in men to cause a feeling which has become known as "lust" to occur whenever a man saw a naked woman's breasts; or her legs, or her pubic area, or her ass, or her long hair falling down her shoulders, or her lower back. In fact, evolution has determined that when men see any part of a naked female body, "lust" is the result. As a matter of evolutionary fact, man does not have to see these areas; lust is instantly meaningful if he can just imagine them like you just did! This idea worked better than planned as the lust was generated easily in men and women quickly determined that this feeling could be exploited for their own purposes as they were given "only when it suits us gene". They quickly determined that they could either go along or they could demand some act in return for satisfying this lust. Unfortunately man did not know what to do with this new knowledge. Women quickly evolved a strategy in which they were able to use the "lust" god gave to men to demand something in return for letting man satisfy his need for "lust." This quickly catapulted women to a position of extreme power as their lust genes were very controllable and they could manipulate men to generate an entire societal manner of living. Well almost. Those humans born with a strong LEOP gene set simply evolved to give both sexes the same desires and women lost their superiority in this gene pool. Thus the Intelligent Designer chose to create mankind in a sexually perfect, but somewhat humorous way. The gods then built some prototypes and tested them. But the design was not an easy one. "Why do we need two opposite sex humans again?" asked Accountant god. "Because then humankind can enjoy the pleasures of sexual ecstasy" opined CEO God. "Eventually they can even fight over this sex thing and get moody and argue, sort of like Sarah god and her mate do every 28 days."

The first design that was experimented with utilized two large size fingers, one on each sex. By pressing them together, organisms were transferred which immediately combined and grew into a third appendage. Alas many problems were encountered. Every time the fingers touched, voila, a new being was created. Since it felt good to touch special fingers, a lot of finger touching was going on. Another problem occurred when the newly formed "thing" dropped from the finger tips on to the ground where it was immediately stepped on by one of the clumsy prototype humans. "That won't work," said CEO god. Make another prototype and give the male a dedicated sex organ on his feet. We should then require the male deliberately choose to mate by spreading "something" over another set of "some things" on the female. We should hide the female "something" so men cannot brag about their ability to "cover it" being superior to that ability of their friends. Maybe we can hide it so well that multiple sex partners will never know who "covered it"? Put the female's contribution in a pouch until the 'something" is big enough to survive on the ground" he said. "And make it feel good!"

The second set of prototype humans didn't work so well either. The act of placing one large toe into the other's pouch was creating a very pleasurable feeling and the two humans were creating all these eggs that just overflowed the pouch and spilled onto the ground. Damn! We need to make it more difficult to want to procreate and we need to keep the egg safe" said CEO god.

Eventually CEO god decided that man would be created with a dedicated sex organ, the penis. Note: this was not quite dedicated as accountant god decided that it would useful to excrete urine when it was flaccid, therefore creating "dual use" efficiency. They equipped man with a penis. Of course it was mostly flaccid, but after all man was supposed to work and not just play. They equipped woman with a vagina. Conveniently the penis was shaped to fit the vagina; sort of.

The vagina looks too much like a mouth," the gods said. "Men will misuse it." "I have an idea," said engineer god. "Why don't we put lots of curly hair around it so men won't see its mouth shape and won't think it is a second mouth?" Thus was created the hairy vagina. Efficiency god, not wanting to be outdone, said "We can make multiple uses for this vagina thing too. "We can use it as the excretory organ for pee, we can have the babies come out through it and we can use it to massage the penis!" "Spectacular" said CEO god.

"Damn, this penis design is useless when soft. I can't even get man to stuff it in manually." One cannot help but to wonder why the intelligent designer didn't just make the penis rigid with a bone so that arousal never became an issue in order to propagate the species? Perhaps it would have been too big to comfortably stash between the legs. A wise designer might have put this always erect penis on the arm alongside the hands. Then any female desirous of fertilization would simply snuggle her vagina onto the always erect, always ready penis. This idea could be carried further if the vagina were placed on the female's elbow, then the couple could procreate anywhere at any time. No mess, no fuss no need for "arousal." In addition placing the sex organs on the elbows would have eliminated peeing thru the same organs evolution taught humans to love to lick.

I can solve that by putting a bone in the penis so it will always be ridged," said engineer god. Unfortunately this suggestion did not pan out either as the erect penis was always getting in the way of male to male games and keeping males from being able to sleep comfortably under a blanket. It should be noted that originally when women still attended this council meeting, 10" penile length was the desire of the male gods, but 6" was decided as sufficient by the female gods as 10" would simply be too much of a potentially good idea. "I know, I know" said Engineer god. "We can have the penis get stiff only when lust is present so it will fit the vagina. I will design a system wherein when man is excited, his normally flaccid penis can become engorged with blood and become fully erect and stiff." "We could take the excess blood we would need from his brain," said efficiency god, "after all he doesn't need his brain when engaging in procreation anyway." The gods decided then that removing all of the blood flow from man's brain was very useful and to this day man cannot think a cogent thought when fully aroused, except when he is frightened, tired or distracted by other females.

So far the intelligent design gods were on a roll of successes. However, Environmental god was still concerned that there would be far too many humans if all the sex led to still more humans. So the council pondered this new problem. "I know" said accountant god. "Let's make the act of fertilization a gambling event of Las Vegas magnitude. Let's give the female only a few eggs for her entire life. Then let's make them travel down a treacherous path to the uterus only periodically. That way should the males ever just engage in the sex act for pleasure, the chances of them creating a fertilized egg would be nil. Actually 1:1,000,000," said Accountant god. "Further following that reasoning" said efficiency god, "the journey for the sperm to finally meet with a female egg can also be highly treacherous, thus ensuring that the chances for fertilization are even less good." "The males will never know for sure which of them actually fathered any offspring," laughed accountant god. Thus the Intelligent Designers deliberately made the sperm swim an Olympic competition. "This journey is so difficult that only the sperm equivalents of Michael Phelps could possibly make it. Just think, he said, "If every sperm has to swim until exhausted, then only 1 out of a million could possibly make it thru to the egg which only possibly might be there. This way only the fittest will survive" he smirked. "But isn't that what evolution would do to ensure survival of the fittest" asked German god. "Well not quite, dick old boy" said CEO god. Why if we designed man to have one gigantic sperm that could swim for miles we would then have to choose which genes we inserted every shot, a whole new set of decisions that we just don't have the time for. Let the million sperm method prevail with 100% randomness except for the LEOP gene set," he opined.

The designer gods then placed the sperm production in two sacs, exposed to the weather and danger by hanging the sacs between the male legs. They then discovered that the sperm could not swim the entire length of an erect penis to enter a female's vagina without some help because the penile canal was so very dry except during urination and the sperm did not have feet, surely a design oversight. So the gods added a seminal fluid generator but couldn't quite find the room for it near the base of the penis or near the sperm generator, so they were forced to locate it far to the rear of the penis. Because the flow of seminal fluid was so essential to the sperm's swim, they also put the seminal fluid feed tube adjacent to the urinary exit tubes. This ensured that when the prostate gland (another chapter? Not!) eventually grew enlarged it would restrict the urinary flow. This in turn required surgery to correct which in turn would terminate the flow of seminal fluid. From the standpoint of evolution this set of useless conditions produced the right effect, effectively causing the sperm to die off before they could impregnate a woman. Since it only occurred in older men, it further ensured that only the young and strong could procreate. Clever design by the god committee! Why, a skeptic might ask, didn't they put the seminal fluid next to the scrotum and just terminate sperm production after the age of 18? This would have enabled men to enjoy the pleasure of sex until they died.

CHAPTER 8

· · · · · · · · · · · ·

Stardust . . .

IT IS STRANGE THAT WE commonly use the expression "The Devil is in the details" which is the same as saying "God is not in the details" when we refer to the complexities of things. If indeed the Devil is in the details of human design, then who is the ultimate designer? Could it be the devil? If it were the devil and not god that created humans, many of the present day attributes of humankind would make more sense. For instance, what if all of animal kind had sex not for procreation but for the pure joy of sex? We could then all use the excuse, "The devil made me do it." Does this theory not also explain the nasty body odors and the loss of body beauty as we age?

If the devil and not the multitude of gods made humankind, then an awful lot of human traits can be explained. For instance, the male need for continuing sexual release without the need for any feelings associated with the act of sex is now explainable as is the proximity of sex and excretory organs. The fact that humankind is continually at war with one another over "religion" makes perfect sense if the devil was the designer. So is the fact that heterosexual sex leading to multiple pregnancies leading to global over population and starvation and hardship is also explainable. Think about the cruelty of the joke on mankind where men lose their hair and women gain hip weight soon after adulthood, now making imminent sense. Alas, we digress. This book and research is all about evolution versus intelligent design.

In our exhaustive research we have uncovered many pseudo scientific facts about the origins of creation and of our own lives. Yet these facts do not lead us to a full understanding of the many mysteries that we have also uncovered and which are still to be researched. These results indeed point to the conundrum that humankind appears to have evolved rather than having been "designed" by the omnipotent gods.

Fortunately more excavation work in the Tibetan highlands has resulted in still more scrolls and with them increasing knowledge of creation. Dr. Daahs and Dr. Fükuppe have returned to the area of the Lost Scrolls of Creation and worked for the last several months to decipher still more of these great mysteries.

In working with his priestly highness Hau Tao, we have analyzed many more scrolls. Fortunately Dr. Fükuppe's ability to rapidly enlarge these images has accelerated our analysis. The findings are quite disturbing as these documents point to a new development in the god like origins of humankind. According to these ancient writings, and paramount among these, is one in which new, female alien gods, take over and complete the process started by the male gods. One scroll also describes their homeland and their way of life.

Apparently, these female gods live on the giant moon "MooonOvrmiamii" located in the galaxy known as "Frfrawhey." They describe a life style largely without male interaction as the male gods apparently live on a different, far messier planet. These female gods have demonstrated great scientific prowess as they were able to design their own spaceship equipped with lots of mirrors and bathrooms on board. They used this craft to follow the male gods during the male god tour of the known universe in which they would seek new golf courses and places to create new life. The male gods would explore other civilizations and ultimately find the paradise of the hereafter. They travelled over 75,000,000 light years to get to Earth. Earth was chosen because the female gods were complaining of the long journey and the lack of specific direction and destination being followed by the male gods. Comments were made suggesting that the male god tour "appears to be mindless and endless" according to written documentation attributed to Queen Asmodeus, apparent leader of the female gods. She was quoted further as railing against the endless time warps from black hole to parallel universe through another black hole to another universe as just boring. "After you have seen one black hole, they all begin to look alike. We could have asked for directions and got here in just less than 20,000,000 light years," she said.

It is from Queen Asmodeus' writings that we have discovered that it was the female gods (goddesses?) whom ended up with the predominance of influence over the design of world that was to be left behind and became Earth. By the time of arrival of the Queen and her ilk, the male gods had become totally bored and had left to explore other space remnants. They took with them the toys they had brought with them like four wheel "GoVroom" vehicles and even had their spiked shoes to enable any sport that another planet may have permitted. "We have done the hard job of designing the bodies and minds these humans need. Lets return in, oh say, the Earth year 2012, and see how humans have progressed?" said CEO god. "Let Queen Asmodeus and her design crew deal with what is left behind," gloated German god. The scrolls left behind by the female gods are extensive. Over 2400 have been found thus far. They detail the actual conversations between the gods as if a court reporter had captured every word.

The scrolls tell us that the final stage of humankind came to be left to the wishes and desires of Queen Asmodeus. The Queen's companions were all of the same mind believing that humans should just be another random species designed to provide sadistic pleasure to the gods. "If we are to leave a civilization behind, let's at least have some fun with them so when we return we can see if they coped any better than we immortal gods would have. Maybe they will evolve and we can learn from them. Remember ladies, the Devil himself lives in the details!" she writes.

"For instance," she said, "Let's have many different varieties of humans." "We can make some white skinned, so that they can't stay outdoors in the sun long without burning from the sun's rays. Others could be brown skinned so they could tolerate the sun and hide better at night. We could make their eyes whiter than normal so they couldn't hide as well! Still others could have slanty eyes and foreboding visages. We could control all of these qualities through the codex we insert in each human those other god's attempted to design. We can call it the "Genomistastic". We could program all sorts of weirdness's and by the random combining of the components of the "Genomistastic" humans can create all kinds of variety. Even we won't recognize the end result. They could actually change to adapt to their environment without us gods tweaking and approving every change, sort of like an evolutionary thing. In fact let's design a gene into the Genomistastic that actually causes random mutations over each generation. That way whenever two humans do succeed in reproduction, the result will be entirely randomized. Who knows what that may produce over time," smirked number two ranking Queen Diabla.

"We could create a lot of chaos and see how they evolve to handle it. For instance, let's ensure that when humans learn to drive, we can omit the driving gene from all persons with slanty Asian eyes. Wow! Imagine an Asian person driving on a crowded highway! We won't allow the two genes for eyes and driving to simply exist together! This is going to be fun. Let's give the Italians the "drive fast" gene but totally delete any rule following genes in them to add to the chaos" said Queen Diabla. "We should give them good taste genes to compensate for their lack of organizational skills" offered assistant queen, Sophia. "And we can give the Germans a set of genes that won't allow them to tolerate disorder of any kind. We can give the British genes that cause their teeth to be at angles and to rot easily. Oh this is so much fun," she giggled. "Oh that sounds great", said Sarah god. "Then we could program the male Genomistastic to cause males to congregate together among other males that would leave "space" for the remaining female humans to do what they prefer. That would allow the female of the new species to go and plan for how they are going to rule the new world," said Oprah god. Now they were all getting into the great scheme that was being concocted. Queen Asmodeus commanded that the lowly female gods search the hinterland for foreskins so that a record of their exploits would be written down and left for humankind to discover. "If you don't find any, just cut them off of every male you come upon," she jokingly directed.

"Let's design the codex to give each human a gene that causes each type of human to congregate together. We can do this even for entire species of fauna. For instance, all birds of a certain feather can always flock together. Humans will then only feel comfortable with other humans that look or talk alike! We can even put a gene into their Genomistastic that will cause them to want to fight other humans that don't look like them, or talk like them," said Queen Diabla. "Don't you worry that they might evolve and want to kill other humans that don't look like them? Ask Queen Oprah. "Hell no," said Queen Asmodeus, "Why would anything kill each other over something as trite as they way they look? Besides, who cares," she said. "But then they won't want to honor us gods that designed them? Queen Diabla said. "What if they find these scrolls?" She asked again.

"Let's give them the choice to choose mates. We could insert a sequence into the codex that lets them choose the body type that they believe will give them sexual satisfaction," laughed Queen Asmodeus. We can have lots of fun with that one. We can let them choose a mate when they are young and good looking. Then we can program their bodies to slowly change and decay over time. They will gain weight and lose all of the body physique that they chose on in their youth," she laughed very hard. "I have an even better idea, she roared; let's insert a gene that makes some of them choose a human of the same sex!!!! Good looking males who would normally be desired by females could opt for another male instead. Females can easily choose another female and that would drive the male gods crazy!" Queen Diabla joined in the raucous laughter. "Just think, all the time those male gods spent on worrying about the size, shape and function of the matching sex organs, all gone to waste," she laughed. Then Queen Diabla came up with the most nasty, nefarious plot of all. "Let's make the female bodies go thru a period of time each year when they are most fertile and lets have their Genomistastic give them a hormonal imbalance that will cause them to be very, very aggressive, and intolerant of males! Think of the chaos it will cause," she said. "Great idea", said Queen Asmodeus, only why make it once a year, make it every, say when does that moon of theirs come to view? Yeah, make it once a moon cycle"! She roared with laughter.

Finally, let's make then all insecure so that they are imbued with a constant need for understanding what they are on Earth for. "Oh, oh" cried Queen Diabla. "Let's make them yearn for an eternal life like we have. Then they could all die without having ever found the answers and they will hate the male gods," she giggled. "Better yet, said Queen Diabla, let's give them a need for believing that an all knowing god made them. They will search forever for the absolute truth and when they find these scrolls, they will be pissed," she laughed uncontrollably.

Well dear reader, Dr. Fükuppe and I are still not finished with our research. These latest revelations are simply mind boggling and we need a rest from our work. We believe that the treasure trove of pseudo science data is inexhaustible. Dr. Fükuppe swears to continue her research into the "divine ratio" and she will spend the next several years measuring bodies and reporting on her results. I will continue my deep hypnotic trances, in addition to helping Dr. Fükuppe with her continuing work as she needs data from both sexes, and as we learn more we will publish another book to keep you on the path to the Ultimate Pseudo Truth.

www.ingramcontent.com/pod-product-compliance
Lightning Source LLC
Chambersburg PA
CBHW021252280526
45784CB00005B/2346